AHEAD

LO OK

classroom
COURSE

展望未来英语教程

（英汉双语简体字版）

intermediate
中级教程

TEACHER'S BOOK
教师用书

3

MADELEINE DU VIVIER

ANDY HOPKINS

JOCELYN POTTER

龚龙生改编

本教程为下列机构的合作结晶：

BBC English　英国广播公司

The British Council　英国文化委员会

University of Cambridge Local
Examinations Syndicate (UCLES)　剑桥大学地方考试
管理委员会

Longman ELT　朗文英语教学部

with the cooperation of the
Council of Europe　欧洲委员会
（协助机构）

上海外语教育出版社
外教社 SHANGHAI FOREIGN LANGUAGE EDUCATION PRESS

LONGMAN 朗文

说明

　　《展望未来》英语教程（Look Ahead classroom course）　是在欧洲委员会的协助下，由英国广播公司、英国文化委员会、剑桥大学地方考试管理委员会、朗文英语教学部等机构合作编辑出版的。与同名的电视教程不同，本套教程着重课堂教学。全套教程分四个等级，每级分教师用书、学生用书和练习册。为适合中国学生学习外语的特点，双语版作了一些必要的改编，使该套教程既适合课堂教学，也适合学生和英语爱好者自学之用。

　　《展望未来·3》中级教程教师用书改编本结合中国学生学习外语的特点和一般英语教学的具体情况，将原书中对中国教师来说不太必要和实用意义不大的内容作了删节，但仍保留原书的课文背景知识注释和有关的语法注释，并将这两类注释作了汉译。此外，对各课中的语法重点，增加了注释性讲解，以供教师参考，也方便学习者自学。改编后的教师用书，主要是提供学生用书与练习册中的练习答案和所有录音的文字内容，让教师在评改学生练习时有所依循，自学者也可据此判别自己所做练习的正误。

　　原教师用书中的 Introduction（序言），详尽地介绍了 Look Ahead 全套书的内容、特点以及有关的教学法；书末的 Teacher development tasks　给教师提供了有关的进修材料。本书对这两部分保留了原文，供教学者参考。此外，结合中国学生的实际需要，删去原刊于书末的各课词汇表，将此词汇表改放在学生用书的书末，并加中文释义。同时对各课中有关的难句，均作了注解，放在词汇表后，以帮助学生自学。这里需要说明的一点是，本套书原有的录像带，暂未随改编本出版，今后将视需要再作考虑。

　　改编本难免有疏漏及考虑不周之处，恳请广大读者批评指正。

Contents

READING/WRITING	LISTENING/SPEAKING	COMPARING CULTURES
R: Quotes describing people's memories R: An article about an unusual lifestyle w: Sentences describing changes w: Gap-fill description of a picture	L: A monologue on how an area has changed L: Short conversations L: A description of a picture s: Giving opinions and reasons s: Role play: meeting new people s: A description of a picture s: Stress in compound nouns	Housing
R: An article about television viewing habits R: A text from a brochure R: A literary extract w: A description of a scene w: Making comparisons	L: Sound sequences L: A conversation about a ride at a film studio s: Agreeing and disagreeing in conversation s: Description of sensations	Postcards
R: An article about borders R: Holiday postcards w: A postcard	L: Brief monologues describing location L: Extracts from conversations connected with air travel L: Conversations at customs s: Describing location s: Discussion: airport security s: Role play: at customs s: Stress and intonation of orders and requests	
R: A report of a crime R: An introductory text about a famous person w: A report of a crime	L: Accounts of crimes L: A monologue by a woman police officer L: A narrative about an incident on patrol L: A story s: Describing a crime s: A short talk	Theft and punishment
R: An article about supermarket sales techniques R: Letters, a memo, an invoice, an order form w: A description of a process w: Letters w: Form completion	L: Extracts from an interview with a car manufacturer s: Describing a process s: Role play: selling a computer s: Discussion: sending a personalised gift s: Sentence stress with *both ... and, neither ... nor, all, none*	
R: An article about small children and fitness R: A literary extract w: Continuing a story creating a mood	L: A medical consultation L: A monologue by a dentist L: Sounds: matching sounds and verbs L: A sound sequence s: Role play: persuading someone to stop smoking s: Role play: calming and advising someone s: Intonation of calming and reassuring someone	At the doctor's Words describing sounds
R: An article about an unusual hotel R: A literary extract	L: Interviews about a ranch hotel L: Dialogue at a hotel reception L: An anecdote about camping s: Asking questions about vocabulary s: Discussion: a holiday at a ranch hotel s: Role play: situations at a hotel s: Telling and retelling stories s: Expressing satisfaction and sarcasm	
R: Case studies of people in different jobs R: An article about a change of career R: Job advertisements w: A brief biography	L: An interview with an art student L: A job interview s: Job interviews s: Using intonation to introduce a new topic	Education

	STUDENTS' BOOK UNIT TITLE	VOCABULARY: TOPICS/ DEVELOPMENT	GRAMMAR	FUNCTIONS
9	**Skin deep**	T: People's appearances T: Models and model agencies T: Ideas of beauty D: Adjectives describing people D: Compound adjectives and prepositional phrases	*Look* + adjective *Look like* + noun *Must, can't, might, could* + *be*	Certainty and probability in the future Guessing Making deductions
	Progress check			
10	**Showtime**	T: Puppets T: Theatre T: Circuses T: Animal performers D: Related words D: Negative prefixes	*Be able* + *to* + infinitive *Manage* + *to* + infinitive *Can* and *could*	Talking about achievement Making requests/deductions Expressing ability/possibility Giving permission Talking about prohibition Giving opinions
	Talkback			
11	**Looking forward**	T: Energy sources T: Life in the future T: Life in a closed ecosystem T: Genetic engineering D: The prefix *self-* D: Compound adjectives (time and size)	*Will* + (adverb) + infinitive *Going to* + infinitive *May/might* + infinitive Articles	Making decisions, promises, predictions Expressing plans and intentions Asking for explanations Introducing examples Interrupting Talking about certainty, probability, possibility, plans and ideas
	Progress check			
12	**News and views**	T: Newspapers T: Objectivity and bias in newspapers T: Reading habits T: Perspectives on events D: Adjectives and their connotations	Past simple passive Present perfect passive	Asking for and giving directions Describing location/people Interpreting/inferring from pictures Expressing opinions and feelings Hesitating
	Talkback			
13	**On show**	T: Exhibitions and museums T: Art T: Children and childhood D: *Go* + adjective D: Words with similar and opposite meanings	Past perfect simple	Telephoning Use of tenses in sections of a report
	Progress check			
14	**In touch**	T: Couriers T: Emergency services T: Home computers D: Verbs for reporting results	Reported speech	Asking politely
	Talkback			
15	**A change of scene**	T: Preparations for travel T: Working holidays/Holidays T: Persuasion T: Travel brochures D: Adjectives: degrees of intensity	First conditional Conjunctions: *if, unless, when, as soon as* Indirect questions Embedded questions	
	Progress check			

READING/WRITING	LISTENING/SPEAKING	COMPARING CULTURES
R: An article about a model agency W: A short report about people's views on changing their appearance W: Descriptions of people	L: A discussion about changing your appearance L: A monologue about running a model agency L: Sound sequences S: Discussion: changing one's appearance S: Running words together	Clothing for different occasions Ideas of beauty
R: A text about puppets R: A literary extract R: An article about circus animals R: A letter to a newspaper W: A letter to a newspaper giving an opinion	L: An interview with the manager of a theatre L: Monologues giving strong opinions S: Considering possibilities S: Role play: booking tickets for a play S: Discussion: performing animals S: Using stress to show disagreement	Puppets
R: A literary extract R: An article about a biosphere W: Sentences justifying a decision W: A paragraph expressing opinion	L: An interview about generating electricity L: Brief comments about the future L: A radio news item L: A conversation about genetic engineering S: Negotiating a shortlist S: Discussion: freedom and technology in the future S: Using intonation to allow or prevent interruption	
R: Newspaper extracts R: Comments for a survey R: A narrative W: A report about reading habits W: A perspective on events W: A newspaper report	L: A monologue about producing a newspaper L: Short interviews about newspapers S: Discussion: newspapers S: An interview about reading habits S: Interpreting events S: Word stress	Newspapers
R: An article about an unusual exhibit R: An extract from a brochure R: A poem W: Describing an experience	L: Two short talks by museum staff L: A description of a photograph S: A short talk about a museum S: Role play: finding your way around a museum S: Discussion: children S: Describing and responding to a photograph	
R: An article about being a courier R: An advertisement R: An article about the use of home computers R: A report of a survey W: A report of a survey	L: A conversation about a delivery service L: An interview with an emergency service operator L: Phone calls L: A conversation about an accident S: Reporting an experience S: Role play: phone calls S: A report of a conversation S: A survey of home computer use	Telephoning Home computers
R: An article about working holidays R: Persuasive and neutral texts W: Form completion W: A persuasive description	L: A monologue by a travel agent L: A conversation at a travel agent's S: Discussion: advice for foreign students working in your area S: Discussion: holiday destinations S: Role play: checking information about a holiday	Holiday destinations

Introduction

The course

Why is the *Look Ahead* course special?

The Longman *Look Ahead* classroom materials have been produced as a result of a unique collaboration between BBC English, the British Council, the University of Cambridge Local Examinations Syndicate (UCLES) and Longman ELT, with the co-operation of the Council of Europe.

SYLLABUS
The core syllabus for the *Look Ahead* corpus is based on the Council of Europe's revised and extended Waystage and Threshold specifications (Council of Europe Press, 1991), the most comprehensive statement of language learning objectives yet available for the 1990s and the new millennium.

ENGLISH LANGUAGE EXAMINATIONS
New examinations at Waystage level (the Key English Test – KET) and Threshold level (the revised Preliminary English Test – PET) have been devised by UCLES based on these latest specifications.

BROADCAST TELEVISION SERIES
BBC English has used the same Waystage and Threshold specifications to produce a series of sixty television programmes for English language learners. These programmes are accompanied by self-study materials, marketed directly to learners at home by the BBC.

LONGMAN CLASSROOM COURSE
Longman English Language Teaching has produced a four-level classroom course, which takes as its core the Waystage and Threshold specifications. Extracts from the BBC television programmes have been selected according to their appropriacy for classroom use and are available on an optional video cassette which accompanies the Longman classroom materials. These Longman materials form the complete *Look Ahead* course for the classroom.

What levels does the Longman classroom course cover?

The course comprises four levels:

Level 1 Beginner/Elementary
Level 2 Post-elementary/Pre-intermediate
Level 3 Intermediate
Level 4 Upper-intermediate

Look Ahead Level 1 is for students with little or no knowledge of English. *Look Ahead* Level 2 takes students beyond the Council of Europe Waystage level. *Look Ahead Intermediate* and *Upper-intermediate* take students up to and comfortably beyond the Council of Europe Threshold level.

What are the components of the Longman classroom course?

At each level, the course consists of:

- a Students' Book,
- a Workbook,
- a Teacher's Book,
- a set of classroom audio cassettes (Class Cassettes),
- a Workbook audio cassette (Workbook Cassette),
- an optional set of two classroom video cassettes,
- a Video Workbook,
- a Video Teacher's Guide (one guide covers two levels).

Underlying principles of the Longman classroom course

The writing of the *Look Ahead* classroom course has been influenced by the following beliefs about English language learning:
- Learners are intelligent individuals who are already proficient in at least one language.
- Learners want to know what they are learning and why. They also **need** this information in order to become more independent as learners as they progress.

- Learners need to develop at the same time a knowledge of grammar, vocabulary, functional language and communicative skills. Attention to the systems of the language is crucial, but the development of fluency and contextual appropriacy are equally important goals.
- Learning takes place most effectively when learners are actively engaged in the learning process.
- Topics should be interesting, varied and relevant to students' lives.
- Learners need to be provided with every possible opportunity to use new language in contexts which are meaningful to them.
- Cross-cultural understanding is an important aspect of language learning.
- Learners want and need to be able to measure their own progress.
- Learners need resources to help them to continue learning outside the classroom.
- Teachers want materials that take into account all of the above and are presented in a clear, principled manner, but that also allow for flexibility of use.

Key features

What are the key features of Look Ahead Intermediate Students' Book?

A MULTI-SYLLABUS APPROACH
Each unit provides presentation, varied practice and contextualised use of grammar, vocabulary, functional language and skills.

FOCUS NOTES
Each double page includes Focus notes in the left-hand margin. These notes highlight the main areas of topic, grammar, functional language, skills, vocabulary development and phonology presented or practised on that double page. This means that learners have a clear understanding at the beginning of each double page of what their learning objectives are.

DISCOVERING LANGUAGE
The Discovering Language boxes in each unit encourage learners to reflect on a particular area of grammar and to deduce rules from clearly contextualised examples. Learners then have the opportunity to test these rules through guided and freer practice activities.

FOCUS ON FUNCTIONS
Activities under the heading Focus on Functions ask students to consider the meaning and appropriate use of particular phrases and language patterns. Students are presented with contextualised examples and are then asked to use the new language in a different context.

SPEECH PATTERNS
Useful patterns of stress and intonation that occur in spoken texts are highlighted and practised.

A WIDE VARIETY OF TASK TYPES
Tasks encourage students' active engagement in the learning process through activities which involve discovery, problem-solving, language use and creative response.

STIMULATING AND RELEVANT TOPICS
Each unit contains a number of related topics. These topics have been chosen for their general interest and for the useful vocabulary and functional language which they generate. Their exploitation encourages personal involvement as learners are asked to relate the topics to their own experiences and interests.

DEVELOPING VOCABULARY
New vocabulary is taught in different ways, but Developing Vocabulary sections help students to explore patterns and relationships within and between words, and to discover useful generalisations.

COMPARING CULTURES
Regular Comparing Cultures sections allow learners to reflect on the similarities and differences between their own and other cultures. The intention is not to promote the value of particular cultural conventions in one part of the world over any other, but to raise awareness of cultural variety.

DOCUMENTARY
The Documentary feature is another strong cultural element of the classroom course. Recordings of real people talking about their lives and work are presented on the Class Cassettes; background information and supporting photographs are provided in the Students' Book. The British and American people who feature in these sections also appear on the Look Ahead Classroom Videos, which can be played later for consolidation and enrichment through the additional visual element.

SKILLS

The development of appropriate language skills (reading, writing, speaking and listening) is an important aim of *Look Ahead Intermediate*, and students are presented throughout the book with a wide range of text types and strategies for dealing with them. The third double page in each unit, however, focuses particularly on the development of the skills needed for speaking, functional writing or creative writing. In each unit students are helped, through integrated skills work which is carefully staged and guided, to the production of a spoken or written text. Listening and reading skills development feature throughout the materials.

LANGUAGE REFERENCE

These pages are at the end of each unit and include a two-part summary of the key language presented. The first part lists the forms and explains the uses of grammatical structures that are presented or reviewed in the unit. In the second part, language functions are listed with examples.

PROGRESS CHECKS

There are eight Progress Checks, one at the end of every second unit. These are informal tests of the grammatical, functional and vocabulary development areas presented in the previous unit or units. They can be done in class or as homework assignments. They give students and teachers an opportunity to monitor progress and to decide whether remedial work is appropriate before errors become too firmly established. The keys to these are in the Teacher's Book.

TALKBACK

Talkback pages are an opportunity for students to bring all the language they know to a highly communicative spoken activity. These tasks are intended to be treated lightly and usually with humour; the emphasis is on the development of fluent self-expression.

What are the key features of *Look Ahead Intermediate* Workbook?

LANGUAGE FOCUS

A wide range of activities provides further controlled practice of the main grammatical and functional areas presented in the corresponding unit of the Students' Book. At least one task in each unit involves comprehension of a dramatic episode in the lives of two characters who run a small company. These scenes, recorded on the Workbook Cassette, are audio extracts from scenes on the *Look Ahead* Classroom Videos. The videos can be used for further consolidation of language practised in the Students' Book and Workbook.

EXPLORING VOCABULARY

This section begins with a list of key vocabulary from the corresponding unit of the Students' Book. The wordlist is recorded on the Workbook Cassette so that learners can practise their pronunciation as they revise the vocabulary. It is followed by comprehension tasks, which check understanding of meaning and usage, and awareness of important sound and stress patterns, and development tasks which encourage students to relate vocabulary to other words that they know, sometimes by appropriate dictionary use.

HELP YOURSELF

Six Help Yourself sections provide learner development activities, such as *Grammar contrasts* and *Understanding dictionaries*, which encourage students to reflect on language and the language learning process. Their purpose is to help students to make the most of their own learning potential.

SHORT STORIES

Three short stories, each in three parts, encourage extensive reading for pleasure. The final part of each story is recorded on the Workbook Cassette for extensive listening. Activities focus on general reading and listening skills, such as prediction, inference and personal response, rather than on language analysis.

FLEXIBILITY OF USE

Workbook activities can be used in several different ways, depending on the needs of a particular class:

– as follow-up homework,
– as additional individual study activities, either in class or in a self-access centre,
– as additional class activities with students working together.

A full Answer Key to the Workbook and the Workbook Tapescript are provided at the back of this Teacher's Book.

What are the key features of *Look Ahead Intermediate* Teacher's Book?

This Teacher's Book contains the following information and activities:

INTRODUCTION

The introduction describes the principles which underlie the Longman classroom course and contains general notes on suggested methodology and classroom practice.

DETAILED LESSON NOTES

There are detailed teaching notes for each Students' Book unit to help teachers in their lesson preparation. These are organised under a number of clear headings. Focus notes provide a summary of the main teaching points in each lesson. Like the Focus notes in the Students' Book, they highlight key areas of topic, grammar, functional language, skills, vocabulary development and phonology. These notes are followed by a suggested procedure for each Students' Book exercise. Other ideas for presenting new language are offered under the heading *Alternative presentation*. Background notes give additional cultural information to help teachers from different backgrounds respond to their students' questions. Extra practice sections suggest further, optional activities to supplement those in the Students' Book. Tapescripts and keys to the Students' Book exercises are also provided within the Lesson Notes.

TEACHER DEVELOPMENT TASKS

Towards the back of this book is a unique Teacher Development section for teachers. This consists of photocopiable worksheets with accompanying notes. These worksheets invite teachers to explore – alone or in structured teacher development sessions – areas of language and classroom practice which are particularly relevant to the challenges of teaching intermediate students.

What is recorded on the Class Cassettes?

The set of Class Cassettes contains all the dialogues, listening comprehension materials and speechwork activities in the Students' Book. The tapescripts for each unit appear in the Lesson Notes in this Teacher's Book.

What is recorded on the Workbook Cassette?

The Workbook Cassette contains listening comprehension material: extracts from a drama story about the lives of two people working for an advertising agency, and passages relating to the final episode of each short story. It also contains wordlists which enable students to practise their pronunciation by repeating key vocabulary from each unit, exercises focusing on patterns of sound and word stress, and Help Yourself tasks (where appropriate). The Workbook tapescript is included at the back of this Teacher's Book.

What are the features of the Classroom Videos?

The classroom video material is an optional component of the course. We do not assume that every teacher in every institution will have access to a video player, and the audio Class Cassettes provide all the necessary listening input to the classroom materials.

For those who do have access to a video player, the Classroom Videos are a valuable source of enrichment and extension material. The video material consists of fifteen units, each of which corresponds to a unit in the Students' Book. Each unit is about six minutes in length and includes most or all of the following:

- A short presenter's introduction to the general topic of the unit.
- Scenes from an ongoing story about the personal and professional lives of two people who run a newly-formed company called Marsh Advertising. Extracts from many of these conversations are also recorded on the audio Workbook Cassette and relate to tasks in the Language Focus section of each Workbook unit. It is recommended that the video conversations are used to consolidate new language already presented through the Students' Book.
- Mini interviews intended to highlight examples of functional language in use.
- A short cartoon which exemplifies key language points.
- A real-life interview with someone from Britain or the USA, showing scenes from their everyday lives. Extracts from these interviews are also recorded on the audio Class Cassettes and relate to tasks in the main body of each Students' Book unit. We recommend that this part of the video is used for review and consolidation purposes

after completing the Students' Book unit. This will require access to a video recorder for a maximum of one lesson a week.

A full video tapescript and detailed suggestions on how to exploit the video material are contained in each video cassette box. General suggestions for video exploitation are included in the Methodology and Classroom Practice section below. The Video Workbook, however, provides a complete range of activities which can be done while watching the videos in class or in a self-access centre.

What are the features of the Video Workbook?

The Video Workbook contains fifteen units of activities, each relating to the corresponding unit of the *Look Ahead* Classroom Videos (see above). Each unit can be done in one session with video facilities – in class or in a self-access centre – in about an hour of work. The sections of the Video Workbook relate directly to sections of the videos and the first three pages are written in the sequence of the sections as they occur on the video. The fourth page of each unit is optional and contains activities that encourage learners to reflect on and move out from the video material. Typical activities on this page include role play, discussion and cultural comparison.

Although many of the characters from the video extracts occur in the Students' Book or Workbook, use of the Video Workbook will ensure that the videos are fully exploited.

Methodology and classroom practice

Variety and flexibility of approach are crucial if we wish to hold the attention of a class over time. An overview of all the possible teaching techniques is obviously beyond the scope of this short introduction. However, we feel that some explanation of the approaches implicit in *Look Ahead Intermediate*, and some standard procedures for particular activity types, may be helpful.

1 Developing skills

The development of language skills is crucial at intermediate level, and *Look Ahead Intermediate* offers real texts for reading and listening to, as well as stimulating speaking and writing tasks. Grammatical, lexical, functional and phonological knowledge and skills are developed through working with these texts and tasks. In order that contexts and tasks should be realistic, activities that develop individual skills are integrated into a clear sequence that embraces all language skills. A listening or reading text may, for example, be preceded by a discussion activity and lead, via a language focus, to a writing or speaking task. In this way, learners have the opportunity to work in a 'text world' of real language where their understanding and responses are the primary motivating factors, and language work arises in as natural a way as possible.

2 Developing reading and listening

Look Ahead Intermediate contains a wide variety of texts for reading and listening to, including newspaper and magazine articles, letters, signs, literary extracts, radio broadcasts, short stories, interviews, documentaries and informal discussions. Texts have been chosen because they are interesting in their own right, but they offer a range of other features relevant to intermediate learners. Through related activities, students are encouraged to develop awareness of the language and other characteristics of the texts, and of how these are affected by factors such as the purpose of the writer/speaker, the nature and motivation of the intended audience, and the conventions of particular text types.

Longer texts often contain vocabulary and structures which are new to students, but a carefully planned sequence of activities encourages them to understand the main ideas first and to guess the meaning of new language from the context. Most texts have a title or introduction and an illustration; these can help students to speculate about context and content before reading or listening. This prediction stage can be extended if you wish; it is surprising how accurately we can predict not only the general content but also particular vocabulary that is likely to occur and the overall structure of the text.

Texts are accompanied by clear tasks in the Students' Book, and in the Lesson notes in this Teacher's Book, but one possible way of approaching texts for reading and listening is as follows:

- Ask students to think about the title and/or picture, and what these tell them about the text. Ask general questions about the text type, purpose, etc. but do not supply answers, since it is important that students themselves are eventually able to answer the questions correctly by reading or listening.
- If you wish, ask about words that are likely to occur and ask students to justify their guesses.
- Students read or listen, thinking about answers to your general questions and confirming or revising their own predictions. This gives a clear purpose to reading or listening.
- Students can then work alone or together to answer the comprehension questions in the Students' Book. If they work together, an additional benefit will be the need to justify answers to a partner. Before they answer questions on listening passages, give them time to read and think about the questions and then play the cassette a second time.
- After a quick comprehension check, direct students to further activities in the book. You may wish to play key listening passages a third time as students answer the questions. It is also a good idea to play the whole text again at the end so that students have the satisfaction of listening to something they now understand.
- You may then want to analyse the text more closely with your students (see **Presenting new language** and **Expanding vocabulary** below). Note that texts on the third double page of each unit (headed Speaking, Writing and Creative writing) are often presented as 'samples' or 'models' for writing and speaking tasks (see **Developing speaking** and **Developing writing** below).

3 Developing speaking

In *Look Ahead Intermediate* there are three distinct types of speaking activity:

A) Activities that are part of the natural exploration of the topic (such as Getting Started and Comparing Cultures), where the focus is on the sharing of experiences, ideas and knowledge, rather than on the language that students use to express themselves.

B) Activities that focus firmly on particular skills of speaking (giving talks, telling stories, participating in interviews, etc.). Six units end

with a double page headed 'Speaking'. Although these pages involve integrated work on all four skills, their main purpose is to prepare students for the development of a substantial spoken output. The particular areas of speaking skills developed are not only important in their own right; they are also areas that commonly feature in the speaking components of public examinations at this level. Fluency and accuracy are both important, as is the appropriate use of vocabulary, functional language, grammar, speech patterns, and features of discourse or style. The main thread of the double page usually involves:

- activities to introduce the topic and learning aims;
- listening to a text which incorporates features that students should aim for in their own speaking;
- reflection on and analysis of the sample – for useful functional language, discourse features, etc.;
- development of vocabulary that students might need to express their own opinions and ideas;
- the main speaking task. This final stage can be set up and managed in the same way as role plays and discussion. (See below.)

C) Activities that provide opportunities for freer practice of target language (grammar and functions). As with all freer practice activities, these are designed to encourage spoken fluency and successful communication rather than complete accuracy, while at the same time showing how well new language has been absorbed and whether remedial work is needed. The two main task types used in *Look Ahead Intermediate* are structured role play and discussion. Also included are lesson-length speaking activities called Talkback (see **Talkback pages** below).

Role play

Role plays are often information-gap activities. Sometimes they involve pairs of students (A and B) looking at different pieces of information, in which case Student B should be directed to the separate section at the back of the book. Students then exchange information in English without looking at each other's books. More often, students are asked to take a role in a conversation and to act it out. In this case, the information gap is created by the students themselves as the conversation develops. A possible procedure for both types of role play activity is as follows:

- Organise students into pairs and ask first A and then B students to identify themselves. Check that students know what to do and are looking at the correct Students' Book page (if appropriate).
- Ask a pair of good students to demonstrate the activity to the class, or demonstrate it yourself with one student.
- Walk around the class and monitor students while they are doing the activity in pairs. It is better not to interrupt unless they are having real difficulty or they ask for help. Note down any common errors relating to language patterns that have just been taught.
- Stop the activity when most students have finished, ask for feedback, and discuss any problems or mistakes that you or individual students have identified.
- Ask students to change roles and repeat the activity, if you wish.

Discussion

Structured tasks encourage students to give their own opinions, to talk about their own lives and to bring their own perspectives as individuals to bear on the topics and texts. Each pair or group discussion is an opportunity for students to develop communication strategies and to say what they mean, even if they do not have quite the right words or complete control over appropriate structures. One approach is as follows:

- Arrange students into pairs or groups, and make sure that they understand the task.
- Start the discussion by asking a question which focuses attention on one aspect of the topic, and then encourage students to continue the discussion in their groups.
- Monitor students while they are talking, intervening if requested but otherwise noting any important problems. Ask additional questions if the discussion is flagging.
- Stop the discussion when a number of groups have stopped talking.
- Ask a student from one group to tell the class about his/her group's feelings or experiences and encourage other students to add to these.
- Ask if students had any language problems during the discussion, and then point out problems that you noticed. Ask the students if they can solve the problems before you provide solutions for them.
- Finally, encourage students to ask you about your own feelings and experiences, particularly if you have a different perspective that they might find interesting.

Talkback pages

The last page in alternate units is called Talkback. Each page provides a major speaking activity that should involve all members of the class. These activities are discussions, games and stimuli for narrative that involve students working co-operatively in order to solve a problem, reach a decision, justify a view or create an effect (of drama, plausibility, etc.). These tasks are highly communicative and are, above all, intended to be fun. Talkback activities are longer than other speaking activities, so full exploitation with teacher feedback can take a whole lesson. They encourage students to draw on all their resources for language appropriate to carrying out the task. Suggestions for management are made in the Students' Book and in the Lesson notes.

CORRECTING SPOKEN ENGLISH

The extent to which correction is appropriate depends, of course, on the aim of each activity. When new language is being presented and practised, accuracy of form is fundamental and students must also be able to produce the correct sound, stress and intonation patterns associated with the language in context. If you are presenting and checking language orally, you will be aware of any problems immediately and can correct mistakes as they occur. During controlled practice of specific language items, students should be able to correct themselves or each other; if they are unable to do this, the language may need re-presenting.

When, on the other hand, your main objective is to encourage an exchange of information and ideas, accuracy will be less important than the fluent and successful communication needed to complete the task. You will, however, probably not wish to ignore errors completely, particularly if they relate to language that has been taught. Some possible ways of dealing with these errors are as follows:

- Encourage students to ask each other or you about mistakes that they think they have made.
- Make notes on common or important errors that you hear while monitoring speaking tasks. After the activity, draw attention to the problems orally or on the board and ask students if they can correct them.
- Record one pair or group of students doing the task. Then play the cassette back to the class. Pause it from time to time to highlight examples of particularly successful communication but

also to demonstrate significant problems. Allow the original speakers the first attempts at correction; then, if necessary, encourage supportive contributions from other students or correct the mistake yourself.

4 Developing writing

There are three main types of writing tasks in *Look Ahead Intermediate*:

A)'Functional' writing tasks (writing letters, postcards, reports, etc.), which students may well need to carry out in real-life situations, including examinations. Activities help students to construct written texts to conventional guidelines – to produce clear, effective, accurate products.

B) Creative writing tasks, which involve exploring language through a process that includes an imaginative personal response. Although many of these creative writing pages display 'literary' extracts, the purpose is not that students should try to produce 'literary' text but that they should become familiar with the kinds of processes that are crucial to any good writing (e.g. the creation of mood and drama, or the modification of texts for different readers). The emphasis here is on choice (of vocabulary, structure, text type, register) in terms of the effect the writer wants to convey. There is no single correct response to the final writing task; students use their imagination, draw on their entire resource of language and have the opportunity to express in writing what they really want to say.

C) Guided practice of target language items. These written exercises are opportunities for practising the grammatical, functional and vocabulary items that are taught in each unit. They occur mainly in the Workbook and can either be done in class at appropriate points in the unit, or as homework.

MANAGING WRITING TASKS

Teachers are often tempted to view writing tasks as individual activities that take place outside the classroom, partly because of the practical problems caused by some students working faster than others. It is, however, often better for writing to be done, or at least started, in class rather than at home, so that you can monitor the work and provide encouragement. Possible modes for classroom writing are as follows:

- Students write alone.
- Students work in pairs, with one student writing and the other making suggestions, collecting information, using a dictionary to check vocabulary, etc.
- Students work in small groups, with one student writing and the others helping (as above).

One approach to a longer task is as follows:

- Check that students understand the instructions in the Students' Book. (What kind of text is it? How long? What other guidelines are there? Is there a sample/model?)
- Elicit the beginning of a possible text orally, writing it on the board one sentence at a time and asking for improvements.
- Set a time limit for individuals or groups to write their own text.
- When students have finished, ask them to look again at their own work, or to read another piece of work, and to check the overall impact of the text (does it communicate effectively?), the vocabulary (have they chosen the best words?), spelling, etc.
- Collect neat pieces of writing for marking. Ask for heavily edited or untidy scripts to be rewritten and submitted in the next lesson.

MARKING STUDENTS' WRITING

1 **Marking 'functional' writing** (See a) above.) Accuracy is important in most types of 'functional' writing. You may find it useful to show errors without correcting them, so that students can have the satisfaction of improving their own work. This requires use of a marking scheme that everyone understands. You can, for example, underline important errors and write a code in the margin to show the type of error, e.g.

    O for organisation
    G for grammar
    V for vocabulary
    Sp for spelling
    P for punctuation, etc.

Encourage students to improve their texts (to organise them better, to choose more appropriate words, etc.) as well as to correct actual mistakes. Then look at the final text again.

Some teachers feel that it benefits students to be given a formal mark for each piece of writing. This should reflect the desire for students to produce good, effective texts according to clear text conventions (e.g. letter openings and closings). Readers of these texts are concerned solely with the final text, and not the process of producing it. A standard mark sheet might award **two** marks. The first (on a scale of 1–5, say) is an overall impression mark (for communicative effect) based on your feeling about whether the text works (as a letter, as a report, etc.). The second mark is a total of marks for more specific features of the text (organisation, vocabulary, grammar, spelling, punctuation, etc.), with a higher number of possible marks for grammar, say, than punctuation.

Using a standard mark sheet that you give to each student encourages students because it gives information about what is good as well as what is not so good about a piece of work. It also allows you and/or your students to keep a record of their written work.

2 Marking 'creative' writing (See b) above.)
It is, we feel, more important that creative and highly personal writing should be displayed (read aloud, put on the wall, etc.) and therefore given positive value than that it should be graded. If you feel you have to give a grade, we suggest an impression mark based on three main criteria:

- active individual participation in the process leading up to writing;
- degree of co-operation with writing partners;
- the effect of the final text, based on the degree of creative experimentation that the text displays.

3 Marking written exercises (See c) above.)
These should be marked for the accuracy of the language items which are being tested.

5 Expanding vocabulary

There are two distinct vocabulary expansion threads in *Look Ahead Intermediate*.

The first concentrates on understanding and using vocabulary related to the topics in each unit. Exercises in the Students' Book encourage interpretation of these words in the context of listening and reading texts. (See **Developing reading and listening** above.) A quick, lively drill (repetition in chorus and then individually) is helpful to establish sound and stress patterns before the words are used more freely. Exploring Vocabulary sections of the Workbook provide a wide range of activities based on a list of important new words, which is also recorded on cassette. Consider allowing students to do vocabulary practice activities in groups, and ensure that they have access to a good dictionary so that it becomes an automatic and valued

resource. Encourage students to write down additional new words in a way that is both meaningful and accessible to them so that they can refer to and build on their lists later. Quick quizzes, simple crosswords, word boxes and word games will remind students of vocabulary from previous units as the course progresses. Finally, encourage students to experiment freely and without inhibition in activities where the main aims are fluency and successful communication rather than total accuracy. This will help them use new language with confidence.

The second thread, Developing Vocabulary, concentrates on aspects of vocabulary form, meaning and use – such as affixation, sense relations and compounds – that go beyond the topic areas. These exercises can be done in small groups, or you can present them as open class activities. Access to a good dictionary is crucial, as tasks often ask students to generate new words and expressions using a particular pattern.

6 Presenting new language

New language in *Look Ahead Intermediate* is always presented in the context of a reading or listening text. These texts are normally **not** specially constructed vehicles for particular language items, but are genuine texts in their own right in which key language occurs naturally. They should therefore be treated as real texts first and only afterwards as contexts for language work. Procedures for dealing with reading and listening texts have been discussed above (see **Developing reading and listening**). After focusing on the purpose and content of a text, one possible approach is as follows:

- Ask questions that draw attention to the use of the new language (*Is this a formal or informal conversation? What do we say when we ask a friend to do something?*) and then to the form (*Which verb form follows 'Would you mind . . . ?'*).
- Ask students to identify other examples of the new language.
- Move on to the Discovering Language or Focus on Functions activities (see below), or straight into controlled practice activities.

DISCOVERING LANGUAGE BOXES
These boxes draw students' attention to contextualised examples of new language structures. They are designed to assist an inductive (guided discovery) approach to

grammar teaching, encouraging students to reflect on language patterns and to formulate possible rules. Students can work through the activities in pairs or small groups and then feed back to the whole class. Alternatively, you can ask students to read the relevant examples and to formulate aloud possible rules in English. Write the best rule or rules on the board, eliciting improvements where possible. In the early stages of the course you may need to provide clear rules at the end of the discussion, but students can also be referred to the Language Reference section at the end of each unit for more help with form and usage. Discovering Language exercises are always followed by guided and freer practice activities in the Students' Book, and further practice is given in the Workbook. You may like to do these practice exercises in class or set some of them for homework. The Lesson Notes in this book often suggest further practice activities.

FOCUS ON FUNCTIONS

By focusing on functional language in context, these sections draw attention to the use of particular forms of expression for effective and appropriate communication. Guided and freer practice activities allow practice in using this functional language through the unit, and the Talkback pages provide a further opportunity for using it freely and by choice. Focus on Functions exercises are usually best done as small group work.

ALTERNATIVE PRESENTATIONS

Not all teachers will wish to use the particular inductive approaches in the Students' Book all of the time; some may prefer to present new language in their own way and then to use the reading or listening texts in the book for consolidation. For this reason, suggestions for alternatives are given in the Lesson Notes in this book under the heading Alternative Presentation. Another possibility for alternative presentation is to use the Language Reference pages (see below).

THE LANGUAGE REFERENCE PAGES

These are placed towards the end of each unit and include notes and examples for new aspects of grammar (form and use) as well as examples of new functional language. Possible uses for these pages are:

- as a reference resource for students working in an inductive mode with the Discovering Language boxes and Focus on Functions exercises;

- as a learning tool for students to read **before** working with the text material in which the new structures occur;
- as a revision aid for students and as a reminder while they do related Workbook exercises outside class.

7 Handling guided practice

Practice activities in the book follow the presentation of new language. Task types vary, but the aim at this stage is normally accuracy – the correct manipulation of language patterns. You may wish to supplement these activities in class with further tasks from the Workbook, or to add more of your own. It may be beneficial for students to work together so that they can learn from each other, since the practice activities are designed as teaching and not testing tools. You may wish to inject an element of competition into, say, a gap-filling exercise, so that students with correct answers score points for their team. Here is one simple approach to a guided practice activity:

- Remind students of the new language (if the presentation of that language took place in a previous lesson), for example, by asking a pair of students to act out a conversation that includes it.
- Ask students to read the instructions. Check that they have understood by giving them a minute or two to think about the first question and then eliciting the answer, accepting corrections from other students if necessary.
- Tell students to work through the activity in pairs and then turn to a different partner to compare and if necessary amend their answers.
- Ask individuals for answers in an appropriate form for that activity. If the task was the completion of a chart, for example, you may want to draw the chart on the board (or an OHT) and ask individual students to come to the front and complete different boxes.
- Identify any problems that some students still have. Highlight them and draw attention to relevant sections of the Language Reference page at the end of the unit. Make a note to do a further check when the language is recycled later in the unit. Alternatively, ask students to do a related Workbook exercise.

8 The importance of pronunciation, stress and intonation

Students should be intelligible when they speak English both to native speakers of English and to non-native speakers with whom there is no other language in common. Inappropriate speech patterns can lead to misunderstanding. For this reason, pronunciation is an important thread in the *Look Ahead* multi-syllabus.

In the Intermediate Students' Book the emphasis is on stress and intonation above word level. Attention is given, for example, to patterns associated with certain functions (like orders and requests), and patterns that have real discourse value (introducing new topics or interrupting). A useful standard procedure for dealing with the sections headed Speech Patterns is as follows:

- Organise students in pairs and play the relevant extracts from the cassette while they look at the questions in their books.
- Ask students to discuss the meaning of particular features of pronunciation, and what the effect of different patterns would be.
- Take feedback in open class.
- Play the cassette again, stopping for individual and choral repetition of key sentences.
- Move into a freer speaking activity (such as a role play) in which the new pattern is likely to occur.

The different sounds of English, which students should be able to recognise and produce, are presented methodically through *Look Ahead 1* and *2*. In *Look Ahead Intermediate* the pronunciation of new words is given attention in the Workbook, where an alphabetical list for each unit is provided in written form and recorded on the accompanying cassette. The Workbook activities invite students to repeat the words on the cassette – best done individually out of class – and also encourage students to recognise, and group together, words that share similar sound and stress patterns.

It is always useful to do some revision of the pronunciation of new words in the classroom after students have done the Workbook exercises. You can, for example, choose words from the list at random and ask individuals to contextualise them in sentences to show that they understand the meaning, and to allow you to check the pronunciation of the words in context. Another idea is to write a stress pattern on the board (e.g. O o o – excellent) and give pairs of students one minute to list five words with that pattern.

9 Supplementing the coursebook

This course, with its different components, is complete and needs no specific supplementation. Indeed, in some situations teachers may feel that they are unable to cover all the tasks presented in a particular lesson, so the Lesson Notes in this book contain suggestions for omitting exercises or setting them for homework. However, each teacher has an individual style of teaching just as each class has its own needs, and the progression of tasks through the Students' Book does not in any way rule out additional activities which you might want to introduce. Some are suggested in the Lesson Notes. Others might include:

- games and puzzles for all kinds of purposes;
- further practice in reading skills using locally available English language texts (such as articles from English language newspapers) that relate to a unit topic, and using graded readers;
- project work suggested by topics in the book and involving teamwork inside and outside the classroom. An example would be the preparation of an English language guidebook for a local museum, including floor layouts, photographs, explanatory text, and so on;
- formal debates on subjects of interest, with students preparing short talks in support of and against a proposition.

Specific ideas for these and many other activity types can be borrowed from teachers' resource books, or from friends and colleagues.

10 Integrating the videos

While use of the *Look Ahead* Classroom Videos is optional and the course is complete without them, they were developed at the same time as the books and audio cassettes and are linked to them in fundamental ways (through topic, situation, character and language focus). The visual element provides additional layers of richness to Documentary sections in the Students' Book and drama sequences in the Workbook; some parts of the videos (such as the short animations) will be completely new to students. Teachers are advised to use the video material for consolidation **after** completing each Students' Book unit, and the Video Workbook provides activities that exploit and move out from this material. Detailed suggestions for using the videos are given in the Video Teacher's Guide that accompanies the Video Workbooks.

Lesson Notes

Welcome to Look Ahead Intermediate

The aims of these two pages are to familiarise the students with some of the topics and people presented in the book and to stimulate their interest in the subjects they are going to study.

EXERCISE 1

KEY

A works for a newspaper
B emergency services operator
C hotel owner
D police officer
E works for a model agency
F an ambulance
G handcuffs
H printing press
I cards from a model agency
J a room key (hotel)
K the front cover of an information leaflet on crime; its purpose is to inform; it might be read by householders
L a sign showing that the area is for horses only; its purpose is to warn/inform car drivers
M the front cover of a magazine; its purpose is to persuade readers to open the magazine; it would be read by women (usually)
N a sign on an office door; its purpose is to inform people who the occupant of the office is; it might be read by people working in/visiting the office
O part of a street map; its purpose is to illustrate location/directions; it might be used by a taxi driver/visitor to the city/emergency services operator

EXERCISE 2

KEY

A–H–N, B–F–O, C–J–L, D–G–K, E–I–M

EXERCISES 3 AND 4

KEY

A / H / N = News and views
B / F / O = In touch
C / J / L = Away from home
D / G / K = Rights and wrongs
E / I / M = Skin deep

EXERCISE 5

KEY

1 = D   2 = A   3 = C   4 = E   5 = B

TAPESCRIPT

One. CATHY ELLISON
I've been with the Austin Police Department for thirteen-and-a-half years. Currently I'm assigned to crimes against property, the theft division.

Two. IVAN FALLON
Working for *The Sunday Times* is dramatically exciting. You are absolutely at the centre of what are the most exciting events that particular week.

Three. JOHN EGAN
Horseback riding is a very important feature of Rancho Encantado and it's certainly one of the reasons that people come from around the country to, to stay at the ranch.

Four. CHRIS OWEN
I think that there's a great satisfaction in finding someone, grooming them, putting them on the road to success, and watching them grow . . . I mean, the change you see in some of the models over a two or three year period is quite astonishing.

Five. MICHELLE REDFERN
The main aim of my job is to help the public and get them the ambulance there as soon as we can – in the shortest time possible.

1 Patterns of life

Changes

Focus	SKILLS
TOPIC • Changes in surroundings **GRAMMAR** • *Used to* + infinitive	• Reading: people's memories • Listening: a monologue • Writing: about changes • Speaking: giving opinions and reasons

语法知识:

"used to + 动词不定式"这一结构只有过去时，指过去的习惯和状态。如果我们说某人 used to do something，我们的意思是不久以前他习惯于做某事，但现在已不做了。要表达相同的概念但指现在，通常用一般现在时就够了。试比较:

He used to play cards a lot.　　他过去常玩牌。
He plays cards a lot.　　　　　他常玩牌。

GETTING STARTED EXERCISE 1

Background note

In the 1980s there was a new development of houses and offices in East London in the Docklands area, formerly a poor area, along the River Thames. It includes the Canary Wharf tower, which is the tallest building in Europe. In the 1990s, the area was hit by the recession and development stopped. There is an electric railway (The Docklands Light Railway) which connects the area to the City of London.

背景知识注释：

位于泰晤士河畔的伦敦东区曾经是一个贫困地区。但在本世纪80年代，那里的住房和办公用房有了新的发展。这一地区包括有欧洲最高的建筑物（Canary Wharf 塔）。90 年代，由于经济不景气的原因，发展停止了。现在有一条电动火车线路将它与伦敦城连接起来。

KEY

1 A = 1990s   B = 1970s   C = about 1910

🎦 LISTENING EXERCISE 2

KEY

It used to be a busy area with lots of work at the docks; now it has become a commercial area.

TAPESCRIPT

Docklands is an area in the east of London, along the River Thames, where boats used to come in and dock. In the last century and the first half of this century, the area was very busy; there were lots of ships coming in, and lots of work for the people who lived there. But things changed in the 1950s and 60s: the docks closed and industries moved out. Then, in the 1980s there was another great change: new buildings went up in Docklands, and new companies moved in.

READING EXERCISE 3

KEY

A the river   B pleasure   C home   D work

EXERCISE 4

KEY

1 False: the housing was cramped (Text C), children didn't have toys (Text B), the father often had no work (Text D), money was always a problem (Text D).
2 True: people were very friendly with their neighbours, they went to the park with other families on Saturday evenings (Text B).
3 True: it provided work for the men (Text D), it was exciting for people to watch the ships (Texts A and B).

EXERCISE 5

KEY

1 Boats: tugs, police launches, sailing barges, river buses, pleasure boats. River buses and pleasure boats carried members of the public as passengers.
2 Jobs: (a) a conjuror   (b) a stevedore   (c) a foreman

DISCOVERING LANGUAGE EXERCISE 6

Note: Students may be confused by the fact that the *s* is pronounced /s/ in *I used to* /juːstə/ *play football* (as opposed to /z/ in *I used* /juːzdə/ *a bowl*). They may also try to pronounce the final *d* on *used*.

KEY

1 c. It refers to the past.
2 No, it is not probable. *Used to* refers to a past habit.
3 Infinitive without *to*.
4 We never used to have, we used to go out, That used to be, Dad often used to come home, Mum used to cry, Dad didn't use to pay. The two negative forms are: *never used to*, *didn't use to*.
5 Did you use to enjoy life in Docklands?

📖 Documentary

LISTENING EXERCISE 8

KEY

1 row /rəʊ/   2 skilled/pro'fessional   3 'purchase
4 de'serted   5 'spirit   6 pro'fessional/skilled   7 thrive
8 sink   9 e'state

🎦 EXERCISE 9

KEY

1 JOBS: Professional people live there, not people who work on the docks.
   WEEKENDS: Docklands is deserted at weekends because the professional workers only live there during the week and then go to weekend homes in the country.
   COMMUNITY LIFE: There is no social/community life any more because people are not there at weekends.
2 There were whole estates and rows of houses lived in by people who worked on the docks. Some docks had their own social clubs with dances on Saturday nights and parties for children.

TAPESCRIPT

TERRY WARD

The kinds of people and the, the jobs that they used to do in, in the last century in the docks were all relevant to Docklands itself and the industries which were joined to it. Today, of course, the people are completely different and there are a lot of professional people now, skilled accountants, doctors, lawyers. They're the kind of people who would have a town house and then would want to get away to the country, somewhere in the countryside. A lot of the flats and houses that have been purchased in the docks, well at the weekends they're empty. You don't see them. They're only used during the week when the people are working. Therefore they don't really constitute a community, because at the weekend, when the social life used to be at its most active, now it tends to sink and go quiet. And an awful lot of the new docks estates and areas are deserted at weekends. And you've got nothing any more. In fact it's like a ghost town.

When Docklands was really thriving many years ago and they, they used to have whole estates and rows and streets of houses which were all dockers or related industries, all the people lived there, they had a very close-knit community spirit. And they certainly used to make all their own entertainment, an awful lot of it; and they used to organise trips – you know, days down to the seaside and things like that. Certain docks used to have their own social clubs, for instance, where they'd have dances on Saturday nights and things, and various parties for the children.

WRITING EXERCISE 10

SUGGESTED KEY

People in Docklands used to live and work on the docks, but now professional people live there. The river used to be very busy, but now there are very few ships. There used to be lots of work on the docks, but the docks closed. People used to live in houses, but now they live in modern blocks of flats or houses. They used to go to the park on Saturday evenings, but now the people who live there go to their country houses at the weekend. There used to be a strong community life, but there isn't any more because people are rarely there at weekends.

Moving on

Focus	
TOPIC	**SKILLS**
• Changes in lifestyles	• Reading: an article
GRAMMAR	• Listening: conversations
• Present simple	• Speaking: role play
• Adverbs of frequency	**VOCABULARY DEVELOPMENT**
• Present progressive for current events	• Compound nouns (noun + noun)
• Present progressive for fixed future plans	**STRESS PATTERNS**
FUNCTIONS	• Stress in compound nouns
• Greetings	
• Asking/talking about health	
• Introductions	
• Responding to introductions	

语法知识:

- **Present simple**

  一般现在时的最普通用法就是指"一般时间"—— 也就是说,表示重复发生,或所有时候或任何时候都发生的动作和情况。

  I go running three times a week. 我一周跑步三次。

  有些动词通常不能用于进行时。这些动词甚至可以用在一般现在时来表示说话时正在发生的暂时情况。

  I like this wine very much. 我非常喜欢这种酒。(不能说: I'm liking ...)

- **Adverbs of frequency**

  表示确切频率的副词(这种副词确切地说明某种事情发生的频率),一般不用于句中,而用于句末。

  Milk is delivered daily. 牛奶天天送。

  表示不确切的频率的副词,可以用于句中。

  We often play bridge on Sunday nights. 我们常常在星期天晚上打桥牌。

- **Present progressive for current events**

  现在进行时的最普通的用法就是表示在说话的时候正在进行中的动作和情况。

  Hurry up! We're all waiting for you. 快点来! 我们都在等你呢。

  现在进行时常常用来表示发展中的或正在改变的情况。

  The weather's getting better and better. 天气越来越好了。

  现在进行时用来谈论暂时的情况。不能用来表示永久性的情况,或经常发生的事情或习惯。

  My sister's living at home for the moment. 我姐姐目前住在家里。

- **Present progressive for fixed future plans**

  现在进行时可用来表示将来发生的事情。

  What are you doing this evening? 你今晚准备干什么?

EXERCISE 3

Background notes

A **mortgage** is a loan to buy a flat or house over 20 to 25 years. In Britain, it is common to get a mortgage from a **building society** or from a bank. **British Telecom** (the company Andrew Weston-Webb used to work for) is the largest telecommunications company in Britain. It used to be a nationalised company and was privatised by the Conservative Government in the 1980s.

This is a real article from **The Observer** (a serious British Sunday newspaper) about a man who had a relatively normal lifestyle with his wife and daughter. After a number of crises in his life, he went to Australia under a false name. Later, he confessed to the authorities and returned to Britain. He now lives alone in a flat provided by the local council.

Note: *skip* is British English. In American English, they use *a dumpster*.

背景知识注释：

抵押贷款是为了购买公寓房间或房屋长达 20 至 25 年的长期贷款。从建屋互助会 (building society) 或银行申请抵押贷款是很普通的事。British Telecom 是英国电讯公司中最大的一家公司。它以前是国有企业。80 年代，保守党执政时，将它私有化了。

这是从严肃的英国星期日报《观察家》中摘选的一篇文章。它讲述了主人公与他的妻子和女儿过着比较正常的生活。在经历了他人生中一系列危机之后，他化名去了澳大利亚。后来，他向当局自首，回到了英国。现在，他独自住在当地市政厅提供的公寓里。

注释： 垃圾车斗，英国人叫 skip，美国人则叫 dumpster。

KEY

skips: (noun) builders' large metal containers for taking rubbish away
'left-over: (adj) (e.g. food) unwanted/not sold
'mortgage /ˈmɔːgɪdʒ/: (noun) money borrowed from a bank or building society to buy a house/flat
'pension /ˈpenʃn/: (noun) an amount of money paid regularly (by the government or a company) to someone who has retired

EXERCISE 4

KEY

1 They come from other people, builders' skips.
2 Because he is concerned about the environment and wasting things unnecessarily.
3 He sells them to other people at street markets.
4 He used to be an executive with British Telecom and used to live in a large house in the suburbs.
5 Yes. He says 'I feel much happier living here. Life's an adventure this way.'
6 Open answers.

DEVELOPING VOCABULARY EXERCISE 5

KEY

1 A compound noun is made up of two words which together make a new noun. In these examples, the compound nouns are all made up of two nouns.
2 *identity* cards   *day*-time   *vegetable* markets
fish hooks/markets   *curtain* hooks
kitchen shelves/chair/hooks   *bed*-time
3 *bookcase*/ club/end
armband/hole/ rest
lamppost/stand
coat hanger/ stand
phone book/ box/ kiosk/
streetcar (American English for *tram*)/life
nightcap/dress/fall/life/light/ porter/ shift/shirt/
watchman

SPEECH PATTERNS EXERCISE 6

KEY AND TAPESCRIPT

In compound nouns, the main stress is generally on the first noun.

bookshelves   armchair   lampshades   night-time
coat hooks   phone cards   street markets

Extra practice

Draw an outline of a room on the board (see below).

Dictate the description to the students and they draw a picture of the room. Then they describe the room to you. Complete the drawing, then write their description on the board as a class and encourage correction of any errors, especially the stress patterns of the compound nouns.

Description

The bed is opposite the door and there is a bookcase next to the bed on its right. Under the window, there is a dining table with a fruit bowl on it. Opposite the window on the wall, there are some bookshelves with an armchair under them. In the middle of the room is a round table with two phone books and a phone card on it. There are three coat hooks on the back of the door. Hanging from one of them is a nightdress. On the left of the door, there is a coat stand.

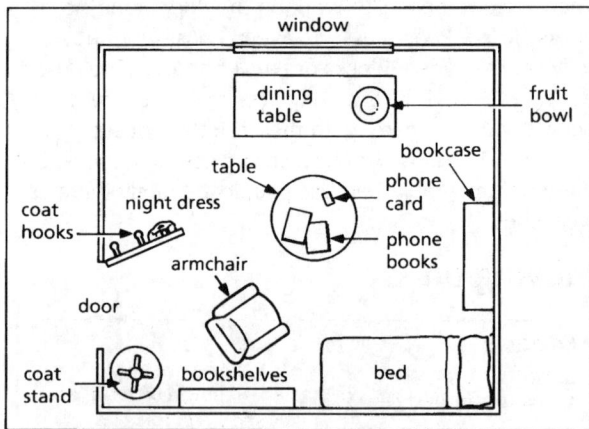

DISCOVERING LANGUAGE EXERCISE 7

KEY

1 a – C   b – A   c – B
2 a – present progressive (*is going down*)
b – present simple (*hardly ever buys/goes out/looks through/sells them/I tell them/pass on/people love it*)
c – present progressive (*is sitting*)

EXERCISE 8

KEY

1 hardly ever
2 always, sometimes, frequently, often, rarely, occasionally

FOCUS ON FUNCTIONS EXERCISE 11

SUGGESTED KEY

1 Open answers.	
2 Formal greetings:	Good morning/afternoon/evening/night.
Informal greetings:	Hello. Hi. How are things? How are you?
Formal introductions:	I'd like to introduce you to ... Can/May I introduce you to ...? How do you do? I'm pleased to meet you.
Informal introductions:	Hello, my name's ... This is ..., Nice to meet you.

🔲 LISTENING EXERCISE 12

KEY

1 The first one is informal. The second one is formal.
2 open answers

TAPESCRIPT

One.

JULIE:	Hi, Madeleine! Good to see you again. How are you?
MADELEINE:	Hello, Julie . Fine, thanks. And you?
JULIE:	Oh, fine. How's Charlie?
MADELEINE:	Oh, he's well. Anyway, Julie, this is Dan. He's working with us today.
JULIE:	Hi, Dan.
DAN:	Hi. Nice to meet you.

Two.

WOMAN:	Julie, let me introduce you to Signor Bettinelli. Signor Bettinelli is the mayor.
MAYOR:	Good evening. How do you do?
JULIE:	How do you do?
DIRECTOR:	And Signora Bettinelli, this is Julie Simms.
JULIE:	How do you do?
SIGNORA BETTINELLI:	Pleased to meet you.

SPEAKING: Describing pictures

Focus

TOPIC
• Housing

FUNCTIONS
• Describing location/people/buildings

SKILLS
• Listening: a description
• Speaking: a description
• Writing: gap-filling

GETTING STARTED EXERCISE 1

KEY

1 the picture in the centre

COMPARING CULTURES EXERCISE 2

KEY

1 The picture on the left shows a block of flats.
The picture in the centre shows a cave.
The picture on the right shows houses on stilts.
2 Open answers.

🔲 LISTENING EXERCISE 3

KEY

1 the picture in the middle   2 b, then a   3 b,c,a
He describes them in this order because his main interest is in the cave dwellings. The cave that he describes first is the most important feature of the picture.

🔲 EXERCISE 4

KEY

1 a) painted white, inhabited, cool and pleasant
b) quite poor, wearing casual summer clothes
c) flat plain, plateau, not much vegetation
2 a) the house   b) the painted facade
c) the brown rock   d) the picture   e) this same hill
f) the house   g) the sloping rock
h) this particular rock   i) the two high points
j) the next building
3 It's probably ...   I can't really tell.   I suppose ...
It's quite possible ...   perhaps ...   ... appears to be ...   They're probably ...   I don't imagine ...
... presumably ...   I'm not quite sure ...
... looks like ...   This must be ...   I should think ...

TAPESCRIPT

This is an amazing picture. In the foreground there's a house built into the side of a bare hill. The house is actually cut out of the rock, and the front is painted white. There's a single window with a, a pink curtain across it and a wooden door. At least, it's probably made of wood; I can't really tell.

This rock house is clearly inhabited because in front of the house I can see washing hanging on a washing line in what looks like the front yard. Then, above the painted facade of the house, they've built a chimney ... so there's a white chimney pot on top of the brown rock. They've also put up a television aerial, which is right in the centre of the picture, so I suppose there's electricity inside. It's quite possible that at the back of this same hill there's another door – or the facade of another house, perhaps.

At the side of the house, to the left and below the sloping rock, there's a flat area and the family seems to be using it as a patio. Anyway, they've got chairs there, and one person appears to be serving food. They're wearing casual summer clothes and they're probably quite poor because I don't imagine that the house is very luxurious. Further to the left and below the patio area is another chimney, which presumably belongs to a different, lower cave dwelling.

Em, in the background, a long way from this particular rock, there's another range of hills. Then, between the two points, there's a relatively flat plain – or a plateau, I'm not quite sure – and then there's a small town. It looks like a town because there are a lot of white houses and each one is quite close to the next building.

This must be a hot country because the sky is blue and there isn't much vegetation. I should think the cave dwellings are actually really cool and pleasant to live in.

WRITING EXERCISE 5

SUGGESTED KEY

1 casual   2 probably   3 At the side of/in front of
4 perhaps   5 close   6 on top of   7 Below
8 Between   9 painted   10 sloping

Progress check: Unit 1

Grammar and functions

EXERCISE 1

1 used to study   2 does she do   3 did Tim use to
live ?   4 do he and Sheila live   5 Do they go
6 do they do   7 Did Tim use to go   8 did Sheila use to
go   9 Does she go   10 Did they use to go

EXERCISE 2

1 Sheila didn't use to go to concerts. She used to go to discos and clubs.
2 Sheila and Tim don't live in a flat. They live on a houseboat.
3 They don't work in an office. They work in a school.
4 They didn't use to go on walking holidays. Sheila used to go on holidays abroad. Tim used to go on cycling holidays.
5 Tim doesn't go to concerts. He goes to parties.

EXERCISE 3

1 Sheila used to study Arabic and Persian.
2 Not possible.
3 Tim used to cycle with friends at weekends.
4 He used to go to concerts once or twice a week.
5 Not possible.

EXERCISE 4

1 They are visiting his parents next weekend.
2 They often give parties on the boat.
3 Sheila is teaching at the moment.
4 Tim is not working this morning.
5 They hardly ever go out on weekdays.

EXERCISE 5

1 Hello/hi   2 Fine thanks   3 And you?   4 this is
5 Pleased/Nice to meet you   6 Good morning
7 Let me introduce you to   8 How do you do
9 How do you do

Vocabulary

EXERCISE 6

1 houseboat   2 lifestyle   3 street market
4 film show   5 community spirit   6 river bus
7 insurance company

2 *Pleasures*

An evening at home

Focus	FUNCTIONS
TOPIC	• Stating preferences
• Television viewing	SKILLS
GRAMMAR	• Reading: an article
• Adjective + preposition + noun/-ing	
• *Love, like, enjoy, hate, prefer + -ing*	

语法知识：

love, like, enjoy, hate, prefer 之后可跟 -ing 形式。如
I like walking in the rain.
我喜欢在雨中行走。
常用的此类动词还有 admit, appreciate, avoid, consider, dislike, mention, mind, risk, suggest 等。

READING EXERCISE 2

KEY

1 The Carter family, who watch TV all the time, had to live without one for a month. The Fox family, who have never had a TV, borrowed the Carter's TV for a month.
2 open answers

EXERCISE 3
Background note
MTV is a satellite TV station which plays rock and pop music all the time.

背景知识注释：
MTV 是播放摇滚音乐和流行音乐的卫星电视台。

KEY

	BEFORE THE EXPERIMENT	DURING THE EXPERIMENT
George	ate in front of the TV	ate/talked with the family
Sandra	watched MTV all evening	played games with the family had more time to do her homework and started piano lessons
Angela	ate/talked together as a family	watched a few programmes but had to plan her life around the TV rarely saw children conflicts over homework
Peter	played family games and read	enjoyed watching TV programmes helped with his schoolwork

EXERCISE 4

KEY
1 box (headline)　2 to rush away (George)
3 to miss (George)　4 crazy (Sandra)　5 nag (Angela)
6 depressing (Angela)

DISCOVERING LANGUAGE EXERCISE 5

Note: The expression *-ing (verb) form* is used in *Look Ahead Intermediate* instead of 'gerund' or 'present participle'.

KEY
1&2　interested + in + noun/(an *-ing* verb form)
　　bored + with + an *-ing* verb form/(noun)
　　tired + of + an *-ing* verb form/(noun)
3　*-ing* verb form

FOCUS ON FUNCTIONS EXERCISE 7

KEY
1　a) 2, 3　b) 1
2　1 prefer + . . .*-ing* form

2 'd rather not + infinitive without *to*
3 'd rather + infinitive without *to*
3 would
4 I'd rather not eat at home.
5 No.
6 The expressions are much more polite than just saying *No*.

A day out

Focus	SKILLS
TOPIC • A visit to a film studio GRAMMAR • *So do I.* • *Neither/Nor do I.* FUNCTIONS • Agreeing and disagreeing	• Reading: texts from a brochure • Listening: sound sequences, a conversation • Speaking: agreeing and disagreeing in conversation

语法知识：
so 可以用在"助动词 + 主语"前，意为"也"。需注意倒装语序。如
"I'll have whisky." "So will I."
"我要威士忌。" — "我也要。"
"I like whisky." "So do I."
"我喜欢威士忌。" — "我也喜欢。"
neither 和 nor 也都可以用在句子的开头，表示"也不"。这两个词后面跟倒装词序。在这样的结构中，neither 的意思与nor 并无真正的不同。但在正式文体中 nor 用得较少。
"I can't swim." "Neither can I."
"我不会游泳。" — "我也不会。"
"Jack didn't like the play." "Nor did we."
"杰克不喜欢这出戏。" — "我们也不喜欢。"

GETTING STARTED EXERCISE 1
Background notes
Universal Studios is one of the largest film companies in the USA. The original film studio was in Hollywood, Los Angeles but now there is also one in Florida. It is possible to visit the studios as a tourist and see where and how films are made. **Ghostbusters** was a popular film made in 1984 about an imaginary firm in New York which tries to get rid of ghosts. A sequel was made some years later. **King Kong** was a film made in 1933 (remade in 1976) about a giant gorilla which terrorised the people of New York, taking a young woman prisoner and climbing the Empire state building, holding her under his arm. **Earthquake** (1974) was a popular 'disaster movie' about a huge earthquake which destroys Los Angeles.

背景知识注释:

环球电影制片公司是美国最大的电影公司之一。最初的电影公司建于洛杉矶的好莱坞。现在在佛罗里达州也有一家公司，旅游者可以参观该电影制片公司，可以看到电影是在何处而又如何拍摄的。《巨鬼》是在1984年拍摄的很为流行的影片。此片描述了一家虚构的位于纽约的公司是如何摆脱鬼怪纠缠的故事。几年后，又拍摄了续集。《超级大猩猩》拍摄于1933年，1976年重拍。这部影片讲述一只大猩猩的故事，描述大猩猩是如何劫持了一位年青的妇女，爬上帝国大厦，而使纽约人惊得目瞪口呆。《地震》(1974) 是关于一次强烈地震如何摧毁了洛杉矶的"灾难性影片"。

KEY

1 in Florida, USA　2 films　3 Hollywood
4 Visitors can: go on rides (a), see where films are made (c), and meet actors (d).

SPEAKING EXERCISE 2

KEY

A is from *Ghostbusters*, B is from *Kongfrontation*,
C is from *Earthquake . . . The Big One*.

READING EXERCISE 3

KEY

1　1 *Kongfrontation*　2 *Earthquake*　3 *Ghostbusters*
2　See above.

EXERCISE 4

KEY

fan'tastic – great　'terrifying – very 'frightening
'crazy – wild　in'credible – unbe'lievable
huge /hju:dʒ/ – e'normous　'famous /'feɪməs/ – well-'known ('well-known when before a noun)
'evil /'i:vəl/ – bad and 'dangerous

EXERCISE 5

KEY

1　'smashing /'smæʃɪŋ/(Text 1)
2　'crushing /'krʌʃɪŋ/(Text 1)
3　trapped /'træpt/(Text 1)
4　sur'vive/sə'vaɪv/(Text 1)
5　co'llapses /kə'læpsɪz/(Text 2)
6　'buried /'berɪd/(Text 2)

🎞 Documentary

LISTENING EXERCISE 6

SUGGESTED KEY

Kongfrontation: the sound of King Kong roaring, smashing buildings and cars, people screaming

Ghostbusters:　the sound of guns firing, people shouting
Earthquake:　the sound of water rushing, buildings collapsing, fire, trains

📼 EXERCISE 7

KEY

(Tapescript is sound effects only)
1　*Kongfrontation:* King Kong is roaring, people are screaming, helicopters are flying, people are running
2　*Ghostbusters:*　people are shouting, guns are firing
3　*Earthquake:*　the ground is rumbling, people are screaming, buildings are collapsing, water is rushing

EXERCISE 8

KEY

open answers

📼 EXERCISE 9

KEY

a) Neither do I.　b) Oh, I did.　c) So am I.
d) Nor am I.

TAPESCRIPT

WOMAN: Wow. That was fan*tastic*! Did you like it?
MAN: Oh, yes. It was *great*, wasn't it?
WOMAN: That part where the train was coming towards us. I was terrified!
MAN: So was I. And then that burning oil tanker. All those flames. They were *real*. I mean, it was a *real fire*. I don't know how they do it.
WOMAN: Neither do I. It was amazing. And that last part . . .whew!!
MAN: What? The water, you mean? I didn't like the water.
WOMAN: Oh, I did. It was really frightening.
MAN: Yes, but you know I don't like water. I've always been afraid of it.
WOMAN: But that's the point.
MAN: What do you mean?
WOMAN: Well . . .that's why people go on these rides . . . because they're frightening.
MAN: Huh? Well, I suppose that's right. But that was a bit too frightening for me. Anyway, do you want something to drink? I'm really thirsty.
WOMAN: So am I. There's a restaurant over there.
MAN: Oh, no! Food? After that ride? I'm not interested in food at the moment.
WOMAN: Nor am I . . . let's find a shop and get some drinks.

DISCOVERING LANGUAGE EXERCISE 10

KEY

1 a) So was I.   So am I.
   b) Neither do I.   Nor am I.
2 Oh, I did.
3 The verb comes before the subject when the speaker agrees, e.g. *So can I.* It comes after the subject when the speaker disagrees, e.g. *Oh, I did.*
4 The same verb form is used in the comment and in the response, e.g. *I was terrified!   So was I.*

CREATIVE WRITING: Describing scenes

Focus	VOCABULARY DEVELOPMENT
TOPIC	• Adjectives associated with colour words
• Landscapes	• Verbs referring to light
SKILLS	
• Speaking: description of sensations	
• Reading: a literary extract	
• Writing: a description of a scene, making comparisons	

READING: A LITERARY EXTRACT EXERCISES 2 AND 3

Background notes

Capri is an Italian island near Naples. It is very popular with tourists. **Sainsbury's** is a large chain of British supermarkets, selling mostly food, drink and household goods.

KEY FOR EXERCISE 2

1 It's early evening and the sun is going down.
2 200–300 feet above the sea.
3 Not very windy – there's a gentle breeze.
4 There are lemon trees, pine trees and honeysuckle.

KEY FOR EXERCISE 3

1 No. He has never been to this spot before. He describes himself as a tourist.
2 Yes. He describes the scene in the first person (I).
3 No. The sliver of moon, the pale blue evening sky and dusk suggest that the sun has already set.
4 It is a British supermarket. He refers to 'the household products section of Sainsbury's'.
5 Perhaps, because he relates the smell of the plants to a supermarket, which is an unexpected reference in the context.

EXERCISE 4

KEY

1 sheer /ʃɪə/   2   aquamarine /ˌækwəməˈriːn/
3 jagged /ˈdʒægɪd/   4 faintest /ˈfeɪntɪst/
5 sliver /ˈslɪvə/   6 honeysuckle /ˈhʌnɪsʌkl/

EXERCISE 5

SUGGESTED KEY

a) *see:* breathtaking, never seen anything half as beautiful, spilling down the hillside, twinkling lights, sheer drop, sliver of moon, brilliantly white, hung in a pale blue evening sky
b) *hear:* sea . . . washing against the jagged rocks, the sound of breaking waves
c) *smell:* the scent of lemon, honeysuckle and pine
d) *feel:* I had the feeling that no one had been there for years, a warm breeze pulled gently at my hair

DEVELOPING VOCABULARY EXERCISE 6

KEY

1 deep blue – dark blue   pale blue – light blue
  brilliant blue – bright blue
2 a) rose pink – pale to dark pink
      sea green – dark green
      jet black – deep black
      mud brown – dark brown
      blood red – deep red
  b) open answers
  c) emerald green, grass green, sapphire blue, snow white, canary yellow, lemon yellow, chocolate brown, ruby red, . . .

EXERCISE 7

KEY

NOUNS	VERBS
the sun	shines
diamonds	sparkle
lightning	flashes
stars	twinkle
candles	flicker

Talkback

Working it out

1 The aim of this page is to give the students some speaking practice around the topic of job satisfaction. It gives students the opportunity to practice language of liking and disliking, stating

preferences and agreement/disagreement. First, find out from your students if they have jobs and if they don't, what jobs they would like to do. Then encourage them to discuss in small groups what they like and dislike about the job they have or why they would like to have a particular job. Direct the students to the pictures and ask them to make a list of the things that may be important in a job: company car, helping people, working outside, opportunities to travel, opportunity to run your own business and be your own boss, money, long holidays, comfortable office and a good position.

2 Discuss as a class what the students have on their lists and then ask them to add other things: e.g. responsibility, promotion chances, variety, meeting people, flexitime, interesting work.

3 The students then work on their own and prioritise the features from the most important to the least important.

4 The students compare their lists with their partners and discuss why the things are more or less important. They should express agreement/disagreement, and then justify their reasons.

5 In pairs, they try to think of jobs that would match each list.

6 Finally, each pair reports back to the class about the jobs they would like and why. If you wish, students can write a paragraph about their own job preferences for homework.

3 *Borders*

Lines on a map

> **Focus**
>
> **TOPIC**
> • National borders
>
> **GRAMMAR**
> • *Have to, must, needn't*
>
> **FUNCTIONS**
> • Talking about obligation, absence of obligation, prohibition
>
> **SKILLS**
> • Listening: brief monologues
> • Speaking: describing location
> • Reading: an article

语法知识：

have (got) to 和 must 这两种形式都用于谈论义务。两者的意思不尽相同。谈论某种以谈话的人或听话的人而定的义务，must 用得最多。如果我说你或我 must 做某事，我的意思很可能是说我觉得那样做是必要的。have (got) to 一般用于谈论

某种来自"外界"的义务。如果我说某人 has to 做某事，我的意思很可能是另一个人认为这件事需要做，或者有一条法律，一项规定，一项协议，或诸如此类的东西要求有人做这件事。

I must stop smoking. 我必须戒烟。（我自己想戒烟。）

I've got to see the dentist tomorrow. 我明天得去看牙医。（我同他约好了。）

needn't 用来表示没有义务。

You needn't tell Jennifer — she already knows. 你不用告诉詹妮弗了 — 她已经知道了。

GETTING STARTED EXERCISE 1

KEY

1 Picture B: a river (the Rio Grande)
2 Picture C: mountains (the Andes)
3 Picture A: a desert (the *Rub al Khali* – the Empty Quarter)

🎞 LISTENING EXERCISE 2

KEY

1 1 Spain   2 Sweden   3 Brazil
2 a) to the south   b) in the north
  The preposition *to* suggests a feature is outside the country or area; the preposition *in* suggests inside.
3 east, west, north-west, north-east

TAPESCRIPT

One.
There are mountains in the north which form the border with France. To the south and east of the country is the Mediterranean Sea, and in the west there's a land border with Portugal. To the north-west of the country is the Atlantic Ocean.

Two.
The south of the country is fairly flat with a lot of forest, but there are mountains in the north-west on the border with Norway. The border in the north-east is with Finland, but most of the south and east of the country is surrounded by the Baltic Sea.

Three.
It's a huge country with the Atlantic Ocean to the east. The borders with Peru, Bolivia and Colombia are in the forest areas in the west. A river forms most of the border with Paraguay in the south.

READING EXERCISE 4

Background notes

Liechtenstein is a very small country between Austria and Switzerland with a population of approximately 28,000. **Wiener schnitzel** is a famous Austrian dish which is made of a thin slice of veal covered with breadcrumbs and fried quickly in oil.

背景知识注释:

Liechtenstein （列支敦士登）是位于奥地利和瑞士之间的一个小国家，人口大约 28,000 。Wiener schnitzel 是一道有名气的奥地利菜。它是将涂满面包屑的小牛肉薄片放在油中稍稍一炸做成的。

KEY

The two meanings of *cross* in the title are:
1 to go/pass over a border   2 angry

EXERCISE 6

KEY

He got lost and crossed the border from Liechtenstein into Austria by mistake. He was worried because he did not have his Australian passport or an Austrian visa with him. As he came from the former Soviet Union, he thought he would be in a lot of trouble because of crossing the border illegally.

EXERCISE 8

KEY

1 He used to live in the former Soviet Union.
2 He's got an Australian passport.
3 He left it in his hotel room in Liechtenstein.
4 He became really worried, he froze with fear.
5 Because it used to be illegal to move within or outside the country without permission from the State.
6 Because he is worried that the immigration officials will start asking questions if he appears too confident.
7 Because he thinks he will be treated like a criminal as he doesn't have any documents with him.

EXERCISE 9

KEY

1 border post   2 frontier guards   3 immigration office
4 passports   5 visas   6 stamp

DISCOVERING LANGUAGE EXERCISE 10

KEY

1 infinitive without *to*
2 have to – don't have to   must – mustn't
   The negative of *need* can be *needn't* or *don't need to.*
3 H – I *have to* get to the station.
   I – You *needn't/don't need to* pay for children.
4 H – I *had to* get to the station.
   I – You *didn't have to* pay for children.

FOCUS ON FUNCTIONS EXERCISE 11

KEY

Obligation: Sentences A, C and H
Absence of obligation: Sentences B, D, E, F and I
Prohibition: Sentence G

EXERCISE 12

KEY

1 British passport holders *do not have to/needn't/don't need to* have a visa for the USA.
2 You *mustn't* take Latvian currency out of the country.
3 You *can* pay with American Express cheques.
4 You *must / have to* walk on the right.
5 First class passengers *don't have to/needn't/don't need to* wait.

Coming and going

Focus	**SKILLS**
TOPICS • Air travel • Customs **GRAMMAR** • Reported requests and orders	• Listening: extracts from conversations, conversations at customs • Speaking: discussion, role play **SPEECH PATTERNS** • Stress and intonation of orders and requests

语法知识:

在间接引语中，常常用动词不定式来转达命令、要求、劝告和提议。
I told Tom to be careful crossing the river.
我告诉汤姆过河时要小心。
The old lady downstairs has asked us to be quiet after midnight.
楼下的老太太叫我们午夜过后要安静点。
否定的要求等，要用否定的动词不定式。
Margaret told me not to worry. 玛格丽特叫我不要着急。

GETTING STARTED EXERCISE 1

KEY

1 The cartoonist is commenting on the high level of security at airports today. He/she probably thinks that it is too high.
2 & 3 open answers

LISTENING EXERCISE 2

KEY FOR EXERCISES 2 AND 3

a se'curity guard (2)   b 'customs officer (4)
c 'passport officer (6)   d 'flight attendant (1)
e 'check-in clerk (5)   f 'porter (7)   g 'pilot (3)

EXERCISE 3

TAPESCRIPT

1 Could you put your seat belt on now, sir?
2 Empty your pockets.
3 Ladies and gentlemen, we've just hit a patch of bad weather, so would you please return to your seats and fasten your seat belts.
4 Open the blue case, please.
5 Please don't go into the departure lounge until eleven o'clock.
6 Can I see your passport, please?
7 Could you give me something smaller, sir? I'm afraid I haven't got any change.

EXERCISE 4

KEY

ORDER	POLITE REQUEST
2 3	1
4 5	6 7

DISCOVERING LANGUAGE EXERCISE 5

KEY

1 Dialogue 1 – d) giving an order
  Dialogue 2 – c) making a request
2 giving an order: to tell   making a request: to ask
3 object pronoun
4 infinitive without *to*
5 *to order* Other possibilities are *insist* and *demand*, but these take a different construction (+ *that . . .*).

EXERCISE 6

SUGGESTED KEY

1 He told you to put your seat belt on.
2 She told/ordered you to fill in the form.
3 He asked you to bring him a glass of water.
4 He told/ordered you not to put the bags under the seats.
5 She asked you to open the case.
6 She told/ordered you to stand over there.

Documentary

LISTENING EXERCISE 7

KEY

1 a customs officer with two travellers/passengers
2 at customs
3 in the USA
4 Where have you been? Why? Have you got anything to declare? Did you buy anything there? How long have you been away? Was it your first visit?
5 to declare things that you have bought in another country

EXERCISE 8

KEY

1 I am a US citizen.   2 I reside permanently in the US.
3 currency   4 goods I purchased
5 goods we acquired abroad   6 on reverse side

EXERCISE 9

KEY

TAPESCRIPT

OFFICIAL: Good afternoon, folks. May I see your passports and declaration?

FATHER: Certainly.

OFFICIAL: OK. Let's see, can you tell me where you've been, sir?

FATHER: We've been to England and France.

OFFICIAL: Why? What was the purpose of your trip?

FATHER: Uh ... partly business and partly pleasure.

OFFICIAL: What is, what is your business, sir?

FATHER: I'm in the oil business.

OFFICIAL: OK. Um, I see you're bringing back $500; can you tell me what you're bringing, Carrie?

CARRIE: Just clothes.

OFFICIAL: Oh. Did daddy buy you some stuff?

CARRIE: Yes.

OFFICIAL: Good. Tell me, have you, were you on a farm or on a ranch of any kind?

FATHER: No.

OFFICIAL: OK. Are you bringing back any food items?

FATHER: No, sir.

OFFICIAL: What about any plants, anything like that, animals?

FATHER: No.

OFFICIAL: OK.

📼 EXERCISE 10

KEY

1 a hat, gloves, small handbag, a child's drawing book, perfume
2 the child's drawing book
3 London Gatwick Airport
4 £25.00
5 Because she bought the perfume in Britain where it is duty-free of tax but it is not duty-free in the USA.

TAPESCRIPT

OFFICIAL: Are these all gifts?

TRAVELLER: They are just gifts.

OFFICIAL: OK. Can you tell me what you have here?

TRAVELLER: Certainly. That's a hat and gloves.

OFFICIAL: OK. What about this one?

TRAVELLER: A small handbag.

OFFICIAL: OK. And this one here?

TRAVELLER: A child's drawing book.

OFFICIAL: OK. I'm going to look at it, OK?

TRAVELLER: Mmm.

OFFICIAL: This is all your own personal clothing?

TRAVELLER: Yes, it is.

OFFICIAL: And what is this here?

TRAVELLER: That's some duty-free perfume that I bought back at London Gatwick. I don't have to declare that though, do I?

OFFICIAL: Is it for personal use or is it a gift?

TRAVELLER: It's a gift.

OFFICIAL: Um, I'm afraid so, yes. Did you not include it in your, in your exemption here?

TRAVELLER: No, I didn't.

OFFICIAL: How much did you pay for it?

TRAVELLER: £25.

OFFICIAL: Yes, I'm afraid that you'll have to include it in your declaration, because it's only duty-free from the country you're buying it in, not from the country that you're bringing it into. If you haven't added it to your declaration, you'll have to add it on to it, OK? Why don't you do that right now?

TRAVELLER: OK.

📼 SPEECH PATTERNS EXERCISE 11

KEY AND TAPESCRIPT

1 Could you open your bag, please? Request
2 Could you open your bag, please? Order

SPEAKING EXERCISE 12

KEY

The parcels are: a poster, a video cassette, a bottle of perfume, a shirt.

WRITING: Sending postcards

Focus	VOCABULARY DEVELOPMENT
TOPIC • Holiday postcards SKILLS • Reading: postcards • Writing: a postcard	• Adjectives describing places, feelings and experiences • Adjectives ending in *-ed/-ing*

READING EXERCISE 2

KEY

1 B (because the word England is included in the address)
2 a) Jenny and Sarah are friends.
   b) Ken and Julie are boyfriend/girlfriend or husband/wife.
3 a) both postcards – *Take care* (*Love* if Jenny and Sarah are good friends)
   b) neither postcard – *Yours faithfully/Yours sincerely* (too formal)
   c) Sarah's postcard only – *Regards*
   d) Julie's postcard only – *Love* (unless Jenny and Sarah are good friends)

EXERCISE 3

KEY

1 Both postcards include a, b, d, e, g, h.
2 kisses
3 the omission of *Dear* and of subject pronouns, the verb *to be* and *there is/are*
   Postcard B: Arrived last night after an awful flight –

long delays at the airport. The hotel's very noisy and the food's really disgusting. Having a rotten time – miss you a lot. At least the weather's not too bad. Going down to the coast later today to try to find a quiet beach. Can't wait to see you again. All my love, Julie XXXX

DEVELOPING VOCABULARY EXERCISE 4

KEY

1 1 fantastic   2 huge   3 sandy   4 great
2 1 amazing, dreadful, awful, miserable, wonderful, OK, nice, good, marvellous, exciting, lovely, appalling, superb, fine, tolerable + *time*. It is possible, but colloquial to use *OK* before a noun.
   2 small, long, tiny, immense, enormous
   3 stony, empty, sheltered, rocky, crowded
   4 amazing, dreadful, awful, beautiful, miserable, wonderful, OK, nice, good, not bad, marvellous, lovely, appalling, superb, fine, tolerable
3 Positive feelings: amazing, beautiful, wonderful, OK, nice, good, not bad, marvellous, exciting, lovely, superb, fine
   Negative feelings: dreadful, awful, stony, miserable, appalling, tolerable
   Depends on the context (often neutral): small, empty, long, tiny, sheltered, immense, rocky, enormous, crowded
   Other words: 1 great, terrible, super   2 gigantic, little   3 lonely, golden   4 sunny, disgusting, windy

EXERCISE 5

KEY

1 terrified, disgusted
2 Experiences: exciting, appalling, frightening
   Feelings: bored, interested, disappointed, amazed, relaxed, surprised, depressed
3 a) depressing   b) excited   c) surprised
   d) frightening   e) interesting

WRITING EXERCISE 6

注释：第 28-29 页上的明信片，从左到右分别为：Athens（雅典）、the Seychelles（塞舌尔群岛）Hong Kong（香港）和 Iceland（冰岛）。

Progress check: Units 2–3

Grammar and functions

EXERCISE 1

1 Neither/nor did I.   2 Neither/nor have I.
3 So was I.   4 So am I.   5 Neither/nor can I.

EXERCISE 2

1 She asked me to help her with the housework.
2 She asked me to clean the kitchen.
3 She told me to turn the television off (immediately).
4 She told me not to talk to her like that.
5 She asked me to do the washing-up and not to leave the saucepans for her.

EXERCISE 3

1 A – D; B – C.
2 A: We had to arrive early.
   B: We didn't have to arrive early.
   C: We didn't need to/have to arrive early.
   D: We had to arrive early.

EXERCISE 4

1 I'd rather go to Ireland.   2 I enjoy fishing.
3 Yes, he'd (he would) probably like to come./Yes, he probably would like to come.
4 I'd prefer to drive there.   5 Ken hates flying.

Vocabulary

EXERCISE 5

1 A: terrified/B: terrifying   2 A: tiring/B: tired
3 A: annoying/B: annoyed   4 A: relaxed/B: relaxing
5 A: confusing/B: confused

4 *Rights and wrongs*

Bag snatchers

Focus	
TOPIC	**SKILLS**
• Crime	• Reading: a report
GRAMMAR	• Listening: accounts of crimes
• Past progressive	• Writing: a report
• Past simple	• Speaking: describing a crime
• Conjunctions: *while/when*	**VOCABULARY DEVELOPMENT**
	• Words related to crime

语法知识:

过去进行时最普通的用法是用来描述一个特定的过去时刻正在发生的事情。

When I got up this morning, the sun was shining, the birds were singing. 我今天早上起床时，太阳放光芒，鸟儿在欢唱。

过去进行时不是用来谈论过去多次重复或者习惯性动作的常用时态，这种情况通常要用一般过去时。

When I was a child, we always went to the seaside in August. 我小的时候，八月里我们总是到海边去。

GETTING STARTED EXERCISE 1

KEY

Order of pictures: D – B – C – A

In the second picture, the woman is looking at her watch and one of the boys is approaching her. The woman's husband is opening the car door while her two children are looking in the shop window.

In the third picture, the boy is just snatching the woman's bag from under her arm. The woman is not sure what is happening.

In the fourth picture, the boy is running off with the handbag and the woman's husband is shouting at him. Her two children are looking at the thief.

READING EXERCISE 2

KEY

1 The witness was sitting in a café on Queen Street.
2 The boys were standing in a doorway opposite the café.
3 The woman was in the street, a few metres away from them.
4 She looked at her watch.
5 He snatched her bag and ran off down the street.
6 He started to walk away from the shop.

DISCOVERING LANGUAGE EXERCISE 3

KEY

1 Present progressive (*I was sitting*) describes a situation in progress. Past simple (*I noticed*) describes a sudden action.
2 Past progressive: *they were watching/she was waiting/she was standing/the thief was running.*
Past simple: *she looked/one of the boys snatched/she turned and shouted/he ran off/the other boy started to walk*
The writer is describing an episode in the past. The past progressive provides the background to the past simple actions.
3 *While* is commonly used to introduce a progressive verb form.
When is commonly used to introduce a simple verb form.

EXERCISE 4

KEY

OFFICER: What were you doing when your wife lost her bag?
MAN: I was opening the car door.
OFFICER: Were you watching your wife?
MAN: No, I wasn't. I was facing the car.

EXERCISE 5

SUGGESTED KEY

While the boys were standing in the doorway, the woman looked at her watch.
While she was looking at her watch, one of the boys snatched her bag.
While one boy was running away, the other one started to walk away from the shop.
While the man was opening the car door, the thief approached the woman.

📼 LISTENING EXERCISE 6

KEY

Crime 1 = Picture D   Crime 2 = Picture C
Crime 3 = Picture A   Crime 4 = Picture B

📼 EXERCISE 7

KEY

1 a) *a person*: burglar, pickpocket, thief, arsonist
  b) *a crime*: shoplifting, burglary, theft, vandalism, murder
  c) *something a criminal does*: steal, burgle, rob, mug, murder

2

PERSON	NAME OF CRIME	ACTION VERB
'burglar	'burglary	'burgle
'pickpocket	-	pick 'pockets
'vandal	'vandalism	'vandalise
'arsonist	'arson	set 'fire to
'mugger	'mugging	mug
'robber	'robbery	rob
thief	theft	steal
'shoplifter	'shoplifting	steal
'murderer	'murder	'murder

TAPESCRIPT

One.

PENNY

I was working in the newsagent's – as I do every Saturday. It's always busy on Saturdays and I was the only person serving. I looked up from the till at one point and noticed a very well-dressed woman behaving rather strangely. She was looking at some pens and had a couple of them in her left hand. Her right hand was moving under her coat, as if she was putting something in an inside pocket. It was my first experience of shoplifting, and the problem was I wasn't sure she *was* stealing. The woman looked up and saw me. Then she brought one of the pens over to the till to pay for it. I simply didn't have the courage to ask if there was another one inside her coat, so I just let her leave.

Two.

JONATHAN

I found a burglar in my house one night. It was all a bit strange. It was the middle of the night and I was in bed upstairs. I woke up and thought I heard a noise in the kitchen. We get a lot of burglaries around here, so I thought, 'Oh, no!' I crept downstairs and noticed that light was shining under the kitchen door. I opened the door suddenly and a man ran across the room and out of the back door. On the kitchen table was my video recorder, a pile of CDs, and a half-eaten banana. It's a horrible thought. While I was asleep, that man was calmly stealing my video and eating my food!

Three.

MARY

Nobody has ever burgled my house, but someone robbed me once. They stole my purse while I was watching a football match. I didn't realise until I was on the bus afterwards. Pickpockets are so good at stealing these days, aren't they? You don't feel anything. The worst thing is that when it's happened to you once, everyone looks like a possible thief.

Four.

GEOFF

I was in Los Angeles on the night of the riots. It was really frightening. There was a lot of theft and vandalism. People were smashing shop windows and stealing everything – televisions, computers . . . even washing machines and fridges. And they were just damaging things for fun. They were going into some shops with guns and demanding money from the owners. Arsonists were starting fires . . . gangs of youths were mugging anybody who happened to pass by. I stayed in my hotel room while all this was happening – I was too terrified to go out. The next day I heard there were quite a number of murders too. The whole experience was just awful.

COMPARING CULTURES EXERCISE 10

Background notes

1 **Public embarrassment** means a punishment which physically shows other people that someone has committed a crime, e.g. putting an obvious sticker on a car that is illegally parked.

2 In Britain, it is possible to have a **suspended sentence**, e.g. a six-month sentence, suspended for two years. This means that if the person commits another crime within two years of the first one, they are automatically sent to prison for six months.

背景知识注释:

1. public embarrassment （当众出丑）指的是一种惩罚，它向公众显示某人触犯了法律。例如在非停车处非法停靠的车上贴上显眼的标记。

2. 在英国，有缓期执行这一可能。例如，6个月的徒刑可缓期两年执行。一旦被判缓刑者在这两年中又犯了罪，他就理所当然地被送进监狱囚禁 6 个月。

Women in the front line

Focus	
TOPIC	FUNCTIONS
• Women police officers	• Agreeing and disagreeing with opinions
GRAMMAR	SKILLS
• Reflexive pronouns	• Listening: a monologue, a narrative

语法知识:

当句子的主语和宾语是同一个人时，一般需要用反身代词 myself, yourself, himself, herself, itself, ourselves, yourselves, themselves, oneself.

I cut myself shaving this morning. 今天早上刮胡子时，我把脸刮破了。

反身代词也可与名词连用，来表达"那个人或那个东西而不是别人或别的东西"这一特殊含义。

The manager spoke to me himself. 经理亲自找我谈了话。

反身代词没有所有格形式,我们可以用 own 来表示这个意思。

I'd like to have my own house. 我很想有自己的房子。

GETTING STARTED EXERCISE 1

KEY

In the larger picture several police officers, including a woman, are dressed in 'riot gear', i.e. the clothes and protection used at riots.

In the smaller picture a woman police officer is arresting someone. *The front line* is the place where the fighting/action takes place.

DISCOVERING LANGUAGE EXERCISE 3

KEY

1 a) themselves   b) herself   c) himself
2 *-self* is for singular forms and *-selves* is for plural forms.
3 *I, you, we*: possessive adjectives (*my + self, your + self/selves, our + selves*).
  He, she, it, they: object pronouns (*him + self, her + self, it + self, them + selves*)

EXERCISE 4

KEY

A He's introducing himself.
B It's looking at itself in the mirror.
C She's defending herself.
D They're weighing themselves.

Documentary

LISTENING EXERCISE 5

KEY

3 c)   4 a)   5 c)   6 a)

TAPESCRIPT

CATHY ELLISON

I've been with the Austin Police Department for thirteen-and-a-half years. Currently I'm assigned to crimes against property, the theft division. Theft entails several areas: we investigate white collar crime, we investigate property crimes, we investigate anything to do with embezzlements or any kind of stolen property – we investigate all those types of crimes.

When you graduate from the academy everyone is assigned to patrol; so you have to do at least two years in patrol, and that's what I did. One of the things you do is you learn how to get from one place to the other quickly; you learn what your major streets are; and so you just drive around, trying to find out who lives where and who knows what's going on. Every district has an area where one person knows what's going on, all over.

And I think that people tend to trust people of the same race. And I found that, even when I was taking police calls with another officer who was not a black officer, if the person that we were talking to were black, they would just – for some reason – turn to me and not the other officer.

I get afraid a lot of the times, but one of the things that you're taught in the cadet class is if people don't see that you're frightened, and if you just do your job, and I've talked my way out of a lot of things.

EXERCISE 6

KEY FOR EXERCISE 5

1 b)   2 c)

SUGGESTED KEY FOR EXERCISE 6

currently – at the moment
crimes against property – theft, vandalism, arson, etc.
white collar crime – crimes committed in offices/business e.g. fraud
embezzlements – stealing money which you are responsible for, usually from a business/company
investigate – to find out more information about something

EXERCISE 7

KEY

1 was working   2 received   3 was   4 was hiding
5 drove   6 were talking   7 shouted   8 crept
9 were walking   10 blew   11 pulled   12 shot
13 was

TAPESCRIPT

CATHY ELLISON

Several years ago when I was working patrol in north-west Austin, I received a call of a female who was distraught. She said that someone was in her home. I was the first officer there – I didn't know where my backup was. And she ran to the door and she goes: 'They're in here, they're in here right now.' So, as we were walking through her home, it was dark – her lights weren't on, and she was behind me, you know, just stuck to me like I don't know ... it was just really funny but she was saying that the person was in the bedroom – that's where she saw them. And as we were walking through her apartment, her apartment, the wind came through and blew a curtain right in to me and I thought there was a person that pushed that curtain, and I had my gun drawn and almost shot that curtain.

SPEAKING: A short talk

Focus

TOPIC
- Legendary heroes

SKILLS
- Reading: an introductory text
- Listening: a story
- Speaking: a short talk

VOCABULARY DEVELOPMENT
- Rephrasing more concisely

GETTING STARTED EXERCISE 1

KEY

1 The cowboys and Indians lived in the USA.

2 The Indians were the native inhabitants and lived in tribes. There were many different Indian languages. They lived in wigwams (tents) and hunted with bows and arrows. They were eventually forced to live in special areas called reservations.

3 In American films, the cowboys are usually depicted as the heroes. (Some films today, however, try to give a more balanced view.)

4 open answers

READING EXERCISE 2

KEY

1 Robin Hood lived in a forest in the fourteenth century.

2 He robbed and murdered people in authority. Those he stole from included the Sheriff of Nottingham, rich landowners and members of the church.

3 The poor.

EXERCISE 3

KEY

1 fiction, apparently, (according to the) legend, historical 'detectives' have tried to find evidence, there *was* an outlaw, Robin Hood was a symbol, new stories about him have appeared through the centuries, a legendary figure

2

POSITIVE	NEGATIVE	IT DEPENDS
hero	criminal	rebel
justice		authority
		outlaw

3 Someone who fights against injustice on behalf of a group of people who cannot defend themselves.

EXERCISE 4

SUGGESTED KEY

1 hero, rebel, fighting for justice

2 hero, fighting against authority and injustice

3 criminal, outlaw

DEVELOPING VOCABULARY EXERCISE 5

KEY

1 people who own land

2 people who do not have any/enough money

3 home owners 4 factory owners 5 car owners

6 the rich 7 the famous 8 the young

📼 LISTENING EXERCISE 6

Background note

In former times, **a knight** was a man of noble rank who was trained to fight, especially on horseback. Nowadays, **a knight** is a man who has been given the title *Sir* by the king or queen. **Note:** The narrator of the story uses the present simple for dramatic effect.

背景知识注释：

在历史上，a knight （骑士）为贵族阶层中的一员。他们受过格斗训练，特别是在马上格斗的训练。如今 a knight 指那些受国王或女王爵士封号 (Sir) 的人。

注释：故事的叙述者为了达到戏剧性的效果而采用了一般现在时。

KEY

1 c 2 e 3 b 4 g 5 d 6 f 7 a

📼 EXERCISE 7

KEY

a) time: one day, when, the next day, until, in the meantime, a few weeks later

b) Cause and effect: so, as a result, because

TAPESCRIPT

I'll tell you a story about Robin Hood. One day Robin and his men are walking through the forest on their way back to the their camp when they meet a poor knight. Robin likes having guests so he invites the knight to eat with them. When they finish eating, Robin asks the knight to pay for his meal, because he and his men don't have much money. But the knight explains that he has nothing and in fact he owes a large sum of money to a rich landowner. The reason for this is that the knight's son was in a shooting competition and accidentally killed someone. As a result, the knight has to pay the landowner four hundred pounds or he will lose his land. He asks Robin to lend him the money and Robin agrees. The next day the knight goes to the rich landowner and pays the four hundred pounds. Now he has to repay Robin Hood's loan.

But Robin feels sorry for the knight and has other ideas. He and his men wait on the road until some rich travellers arrive on their way to London. Robin demands money from them but they say that they have little money themselves. 'If that's true,' says Robin, 'then I am sorry for you and I will give you some money for your travels.' He searches their baggage and finds eight hundred pounds. The travellers run away and Robin Hood keeps the money. In the meantime, the poor knight has worked hard to find the money to pay Robin Hood back and a few weeks later, he appears in the forest and gives it to Robin. But Robin says he does not need the money because he already has it, and he gives the knight the other four hundred pounds from the rich travellers. The knight is delighted and rides home to tell his wife the good news.

Talkback

What's your story?

KEY AND TAPESCRIPT

Picture C

I couldn't believe it. It was terrible. It happened two days ago. I was walking along the street . . . I was doing some shopping, and there was a woman in front of me – she was probably about fifty. Anyway, something fell from her bag – just a piece of paper – and I picked it up. She was walking quite fast so I ran after her. When I caught up with her I put my hand on her shoulder to attract her attention. She turned round suddenly and I felt this spray go into my face and . . . well . . . I was terrified. I couldn't see anything. Then someone punched me and I fell onto the ground . . . I didn't know what was going on. Anyway, when it was all over and she realised what was happening, she apologised. The spray was a security thing that some people carry in the States. It's a kind of foam with a coloured dye in it and it leaves its mark for about seven days. The idea is that you can identify someone who has tried to rob you. So . . . I've got to walk around like this for a whole week. People think I'm either crazy or a thief. I don't think it's very funny!

5 *For sale*

The science of shopping

Focus	**SKILLS**
TOPICS • Shopping • Supermarkets • Processes GRAMMAR • Present simple passive	• Reading: an article • Writing: a description of a process • Speaking: describing a process

READING EXERCISE 2

Note: The *s* in *aisle* is silent, /aɪl/.

注释：在单词 aisle 中 s 不发音，/aɪl/.

KEY TO EXERCISE 1

1 The entrance is usually on the left of the building because customers look to their left and move clockwise.
2 It displays fresh fruit and vegetables to give the impression that only healthy food is sold.
3 The basic foods are kept in different places in the supermarket so that customers will have to pass other products to find them, and may buy some of these other goods.
4 Customers move more slowly.
5 At the end of aisles or at eye level.
6 The supermarkets try to keep them full as customers do not like to buy from half-empty shelves because they think there is something wrong with the products that are left on the shelf.

EXERCISE 3

KEY

1 F   2 D   3 E   4 B   5 C   6 H   7 G   8 A

DISCOVERING LANGUAGE EXERCISE 4

KEY

1 the layout is designed, healthy food is sold, they are kept, customers are taken, shoppers are encouraged, supermarkets are paid, sweets are often placed, more is bought
2 *to be* + a past participle
3 Because the position of the sweets is more important than who put them there.

WRITING EXERCISE 5

SUGGESTED KEY

. . . and taken to the packing house. They are weighed and packed and then are put into cold storage. On Day 2, they are driven to Nairobi airport and are flown to London. On Day 3, they are inspected by customs in London and are then transported to the central supermarket store. On Day 4, they are distributed to supermarkets (where they are displayed).

Making a sale

Documentary

LISTENING EXERCISE 1

Background note

The **Morgan Motor Company** is a small family company which makes expensive and exclusive sports cars. The company has deliberately not changed the way it makes the cars for the last 70 years. They value the fact that many of the parts are handmade by their own craftsmen in the small factory in Malvern. There is a seven-year waiting list for each car.

背景知识注释：

摩根汽车公司是一家家庭式的小公司。它专门生产昂贵的赛车。这家公司在过去的 70 年中执意不改变其制造汽车的方式。他们珍惜这一事实，即许多汽车部件都是在位于 Malvern 的一座小工厂里由它的工艺师们手工制作的。要买一辆这样的车，顾客需等上 7 年。

SUGGESTED KEY

1 a Morgan car
2 British, handmade, expensive, luxurious, traditional, attractive, comfortable, reliable

EXERCISES 2 AND 3

KEY FOR EXERCISE 2

1 Charles is his grandson   2 about 100
3 The frame of the car is made of wood (ash) and it is then covered with metal (aluminium or steel). This is how coaches used to be made many years ago.

TAPESCRIPT FOR EXERCISE 2

PRESENTER
Listen to the first part of an interview with Charles Morgan, the Production Manager of the Morgan Motor Company. The company was started by Harry Morgan, his grandfather, and it makes the famous Morgan sports cars. There are only about a hundred workers at the factory; it is small and very traditional.

CHARLES MORGAN

The Morgan is built using a technique which we call coach building. Coach building was very popular in the 1920s and in fact all the best cars in the world were always coach built. And what that means is that, rather like in the olden days with coaches being built out of a wood frame and metal panels, we still build the car that way. We start with the ash tree, and we make a frame, and we then cover it in either steel or aluminium.

KEY FOR EXERCISE 3

1 the customer's name
2 b and d
3 fifty per cent
4 It's unique in its appearance, its character, the way it's built and the fact that each person who works on the car gives something of themselves.

TAPESCRIPT FOR EXERCISE 3

CHARLES MORGAN
The car, once it's finished as a chassis, has a ticket put on it with the customer's name on it; and even at this stage when the car is only two or three days old, the customer can see his particular car going through our factory. The Morgan is built out of a combination of parts which we obtain from external suppliers and parts that we manufacture ourselves. We manufacture many of the major components ourselves because we want to give the car its particular character. The metal parts are made out of a flat sheet of metal in every case. We make, for example, the fuel tank of the car; we make the radiator of the car.

We have agents in approximately fifteen different countries. The proportion of cars that are exported compared to the proportion that are, are sold in the UK is, is half and half; so approximately 50% are, are exported. The Morgan sports car has a unique look, but it also has, I think, a unique character, and that character comes from the materials that are used in the making of the car, and it comes from the, the way it's built, and the people who build it – because undoubtedly of the hundred people who build this car they all put a little bit of themselves into the car.

DISCOVERING LANGUAGE EXERCISE 6

KEY

Both the car *and* the necklace (= 2 objects) *are* handmade.
Neither the necklace *nor* the perfume (= 2 objects) *is* British.
Both of them (= 2 objects) *are* handmade.
Neither of them (= 2 objects) *is* British.
All of them (more than 2 objects) *are* expensive.
None of them (more than 2 objects) *is* cheap.

📼 SPEECH PATTERNS EXERCISE 7

KEY AND TAPESCRIPT

A Both the <u>car</u> and the <u>necklace</u> are hand<u>made</u>.
B <u>Both</u> of them are hand<u>made</u>.
C Neither the <u>necklace</u> nor the <u>per</u>fume is <u>British</u>.
D <u>Neither</u> of them is <u>British</u>.
E <u>All</u> of them are ex<u>pen</u>sive.
F <u>None</u> of them is <u>cheap</u>.
A and C have a different stress pattern.

SPEAKING EXERCISE 9

SUGGESTED KEY

	SOFTWARE	QUALITIES OF THE COMPUTER
teenagers	word-processing games home study music-making	modern, mass produced, fun, attractive, green
parents buying for children	games home study music-making	practical, modern, cheap, mass produced, fun, attractive, reliable
people in business	word-processing design accounts & book-keeping desktop publishing	international, practical, modern, comfortable, reliable, environmentally-friendly

WRITING: Letters

Focus

TOPIC
• Sales

FUNCTIONS
• Letter-writing conventions

SKILLS
• Reading: letters, a memo, an invoice, an order form
• Writing: formal letters, form completion
• Speaking: discussion

GETTING STARTED EXERCISE 1

SUGGESTED KEY

1 She produces T-shirts with slogans/pictures on them.
2 by mail order
3 They are personalised and are specially printed.
4–6 open answers

EXERCISE 3

KEY

1 A and E, C and F
2 Sandra knows the writers of B (addresses her by first name; note form) and F (informal; addresses her by first name)
3 A – from a customer, to order a T-shirt
  B – from Sandra's secretary, to give her a message
  C – from a wholesalers, to request payment
  D – from a customer, to make a complaint
  E – from a customer, to give a message for a T-shirt ordered
  F – from an acquaintance, to apologise for a delay and to accompany C.
4 she was expecting C/F (. . . you asked for)

EXERCISE 4

SUGGESTED KEY

She will probably deal with B first (urgent). She will probably deal with A/E next (to satisfy a new customer), then she will probably deal with the complaint (D), and finally C/F, as these are not urgent.

FOCUS ON FUNCTIONS EXERCISE 5

KEY

1 below it
2 a) 4th b) 2nd c) 1st d) 3rd
3 a) 9th January b) 22nd October c) 1st April
  d) 23rd February
4 Dear (name) + Yours sincerely
  Dear Sir/Madam + Yours faithfully
5 Can you phone him . . . ? (B) Please send . . . (D/E)
  could you please include . . . (E)

READING EXERCISE 6

KEY

a) 4 b) 8 c) 6 d) 1 e) 7 f) 5 g) 3 h) 2

Progress check: Units 4–5

Grammar and functions

EXERCISE 1

1 When did he live? 2 What did he do?
3 No, he didn't. 4 tried 5 wrote 6 Yes, he did.

EXERCISE 2

1 were walking 2 arrested 3 was riding 4 ordered
5 was preparing 6 took out 7 shot
8 was still applauding 9 asked 10 said 11 was

EXERCISE 3

1 are eaten   2 is   3 eat   4 are grown   5 is used
6 is made

EXERCISE 4

1 a) yourself   b) myself
2 a) yourselves   b) ourselves
3 a) themselves   b) herself/himself

EXERCISE 5

1 All of these people are young.
2 Both of the men are wearing jackets.
3 Neither of them has long hair.
4 All of them are drinking.
5 One man is not carrying a bag; both the other people are.
6 None of them is smoking.

Vocabulary

EXERCISE 6

1 stole, theft   2 set, arsonist   3 robbed, mugged

6 Body and mind

Staying well

Focus	SKILLS
TOPIC • Health and fitness **GRAMMAR** • First conditional • Second conditional	• Reading: a magazine article • Speaking: a role play **VOCABULARY DEVELOPMENT** • Prefix *over-* • Suffix *-able*

GETTING STARTED EXERCISE 1

SUGGESTED KEY

You shouldn't smoke. You shouldn't eat red meat and fatty foods. You should take regular exercise. You should get at least six hours' sleep a night. You shouldn't work more than eight hours a day. You should get plenty of fresh air.

EXERCISE 4

KEY

1 Not expressed: ... many adults take regular exercise ...
2 Expressed in the article: ... do the under-fives need exercise routines? ... The author, Lucy Jackson, believes that they do not get enough exercise.
3 Expressed: ... parents should begin exercises with very small babies.

4 Expressed: They would not be overweight if they ate healthier food.
5 Not expressed in the article.

EXERCISE 5

KEY

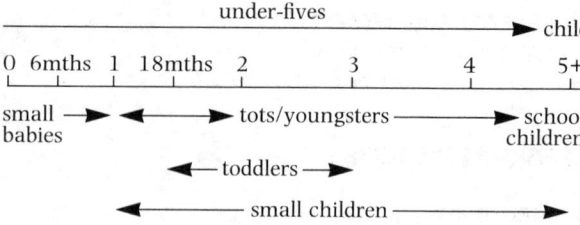

DEVELOPING VOCABULARY EXERCISE 6

KEY

over- = too much. *Overweight* – weighing more than is expected or is usual (e.g. children become overweight)
-able = you can do this. *Disposable* – can be used once and then thrown away (disposed of), *washable* – it can be washed

1 (an evening) you can enjoy
2 (a business) you can make money or get good results from
3 (staff) you can depend on
4 (teachers who) work too much
5 (there are) too many people living in the areas
6 (the bus is) carrying too much

DISCOVERING LANGUAGE EXERCISES 7 AND 8

Notes:

1 Conditional sentences that begin with *If* have a comma after the first clause, e.g. *If it rains, I'll stay at home.* Conditional sentences with the *if* clause second do not have a comma, e.g. *I'd apply for the job if I spoke Italian.*
2 In second conditional sentences, *if* is followed by a past simple form. However, *were* is often considered more correct than *was* in the first and third person singular. In spoken English, the use of *was* instead of *were* is becoming very common, e.g. *If I was you, I'd go on holiday.* It is acceptable in informal discourse.
3 Your students may be reluctant to use the contracted forms of *will* or *would* with subject pronouns, e.g. *I'll, you'll*. They may find the pronunciation of the negative contractions *won't* /wəʊnt/, *wouldn't* /wʊ(də)nt/ and *weren't* /wɜːnt/ particularly difficult and will need drilling.

KEY FOR EXERCISE 7

1 *If* + present simple, + *will*
  If small children *take*
  no regular exercise, bad habits *will continue* . . .
2 *If* clause = a condition main clause = a consequence
3 If toddlers do exercises, they'll stay fit.
4 The first conditional expresses a likely condition.

KEY FOR EXERCISE 8

1 *If* + past simple, + *would* + infinitive (– *to*)
2 *If* clause = a condition
  main clause = a consequence
3 *Were* is used with third person (*the world*) instead of *was*.
4 If there were no cars, children would play in the streets.
5 The second conditional expresses an unreal or unlikely condition.

EXERCISE 9

SUGGESTED KEY

1 If I had a small child, I'd employ a nanny.
2 If children didn't watch television, they'd play with each other.
3 If they played in the streets, they'd get more exercise.
4 If parents didn't listen to experts, they'd worry less about their children.
5 Children would be fit and healthy if they took more exercise.
6 They wouldn't have a bad diet if their parents cooked more often.
7 Teenagers would be healthier if they switched off the television occasionally.
8 If I found my child with a cigarette, I'd explain the dangers of smoking.

Treatment

Focus

TOPICS
- Doctors and dentists
- Medical treatment
- Fear

GRAMMAR
- *Should* + infinitive
- *Ought to* + infinitive

FUNCTIONS
- Giving advice
- Expressing fear
- Calming and reassuring someone

SKILLS
- Listening: a medical consultation, a monologue
- Speaking: role play

VOCABULARY DEVELOPMENT
- Adjective + preposition collocation

SPEECH PATTERNS
- Intonation of calming and reassuring someone

语法知识：

should 和 ought to 的意思很近，它们用来表达职责和义务，提出劝告，即一般说来我们认为人们应该去做的正确的事或好事。请注意，should 和 ought to 可以用来谈论现在和将来，但不能用来谈论过去。要谈那些应该发生而设有发生的事，should 或 ought to 应与动词不定式的完成式连用。
The taxi should/ought to have arrived at 8:30.
出租汽车本来应该在8点30分到。

GETTING STARTED EXERCISE 1

Note: A **pharmacist** (**druggist** in American English) is a trained person who works in a pharmacy dispensing medicines. An **acupuncturist** treats diseases by placing needles in certain parts of the body. A **herbalist** uses herbs to treat illness. A **psychiatrist** is a doctor who is trained to treat people with disorders of the mind. (**Psychologists** are trained in the study or science of the mind. They are not doctors and are not able to prescribe medicines or treatment.)

注释： 药剂师（在美国英语中称为 druggist）受过专门培训，在药房配药。针灸师用金针治病。草药师用草药治病。精神病医师专治精神病。（心理学家指的是对精神、心理作研究的专门人员。他们不是医师，无权开处方或进行治疗。）

KEY

b) If I needed an operation, I'd see a surgeon. /'sɜːdʒn/
c) If I were extremely unhappy, I'd see a psychiatrist. /saɪˈkaɪətrɪst/
d) If I had bad eyesight, I'd see an optician. /ɒpˈtɪʃn/
e) If I wanted natural medicines from plants, I'd see a herbalist. /ˈhɜːbəlɪst/
f) If I wanted treatment with needles, I'd see an acupuncturist. /ˈækjəˌpʌŋktʃərɪst/
g) If I needed medicine for a minor problem, I'd see a pharmacist. /ˈfɑːməsɪst/

EXERCISE 2

Note: *A pain* is a sharp, intense feeling. *An ache* is a continuous, but not violent pain. *-ache* is often combined with *head, stomach, ear, tooth, back*. *Headache* is always a countable noun. The others are often uncountable nouns if they refer to a condition, e.g. *I often have terrible backache.* If you are talking about a single attack, the nouns can be countable or uncountable, e.g. *I've got (a) terrible backache.* In American English, they are more often used as countable nouns. *Painful* and *sore* are both adjectives. They both refer to physical pain, though *painful*, like *pain*, is more intense. *Sore* includes the idea of aching from a wound or infection, e.g. *I've got a sore throat. My legs are sore from doing aerobics yesterday.* The verb *to hurt* can be used for all types of aches and pains.

注释: a pain （痛）指剧烈疼痛的感觉。an ache （疼）指持续性而并不剧烈的疼痛感觉。-ache 常和 head, stomach, ear, tooth, back 等组成复合词。headache （头疼）用作为可数名词。其它的表示状态用作为不可数名词。例如: I often have terrible backache. （我的背常常疼得很厉害。）但如果指一次发作，它们可用作可数名词或不可数名词。例如: I've got (a) terrible backache. 在美式英语中，它们较多地被用作为可数名词。Painful 和 sore 都是形容词，指身体的疼痛。但 painful 像 pain 一样指比较剧烈的疼痛。Sore 包括由伤口、感染等所引起的疼痛。例如: I've got a sore throat. （我喉咙痛。）My legs are sore from doing aerobics yesterday. （由于昨天的健身运动，我的腿很痛。）动词 to hurt 可用来指各种疼痛。

SUGGESTED KEY

A A bee's stung me. It hurts. It's painful. It's sore.
B I've cut myself. I've broken my arm. It's sore.
   It's bleeding. It hurts. It's painful.
C I've got a headache. It aches.
D I've got a pain in my side. It hurts. It's painful.

📼 LISTENING EXERCISE 4

KEY AND TAPESCRIPT

The answers to Exercise 4 are in **bold**.
Part one.
DOCTOR: Good morning. Sit down.
PATIENT: Thank you.
DOCTOR: Now then, what's the problem?
PATIENT: I feel terrible. I feel tired all the time ... and I get headaches.
Part two. (with answers to question 1)
DOCTOR: **How long have you felt like this?**
PATIENT: Oh ... for a few weeks, now.
DOCTOR: **And how old are you now?**
PATIENT: Sixteen.
DOCTOR: **Do you do much exercise?**
PATIENT: Er ... well, no, not really.
DOCTOR: I see. Sports? Cycling? Swimming? Nothing?
PATIENT: Er ... I walk to school. That's about it.
DOCTOR: What about sleeping? **Do you sleep well?**
PATIENT: No, not really. I wake up in the night a lot.
DOCTOR: **Why do you think you wake up?**
PATIENT: Well, I worry a lot – mainly about school work. I'm getting behind and I can't seem to catch up.
DOCTOR: Oh dear. **What time do you usually go to bed?**
PATIENT: Oh, about midnight in the week ... and later at the weekends.
DOCTOR: I see. **And do you drink anything before you go to bed?**
PATIENT: I have a cup of coffee, yes. And I'm always hungry in the evening. I usually have something to eat before I go to bed.
Part 3 (with answers to question 3)

DOCTOR: OK. Well, I'm not surprised you can't sleep. I'm not going to give you anything at the moment. I want you to go home and change a few things about your routine. Then I'll see you again in a few weeks and see how you're doing. OK?
PATIENT: OK.
DOCTOR: Right. First of all, you're probably tired because you're not getting enough sleep. People of your age ought to sleep for at least seven hours a night, all right? So **try to go to bed a little earlier.** Secondly, **you should start to take some exercise.** Go for a short run every day or go swimming. Anything that gives your body some exercise. Right? Then there's the eating. **It's not a good idea to eat just before sleeping and you certainly ought not to drink coffee before going to bed.** Coffee contains caffeine ... and caffeine will keep you awake.
PATIENT: OK.
DOCTOR: **And you shouldn't worry so much.** All this can be solved quite easily. If you do take my advice, you'll sleep better. If you're sleeping well, you won't be tired in the day ... and then your school work will improve.

FOCUS ON FUNCTIONS EXERCISE 5

KEY

1 They give advice.
2 a) *should* is followed by the infinitive without *to*
   b) *ought* is followed by the infinitive with *to*
3 *not* is added, e.g. *should not/shouldn't, ought not to/oughtn't to*

📖 Documentary

📼 LISTENING EXERCISE 7

Note: You can be *nervous* (= rather afraid) before or during an event, e.g. if you have to make a speech in public. A person can have a *nervous* character. *Anxious* usually means worried about something that might happen. *Worried* also expresses anxiety but is a little stronger than *anxious*. *Nervous, anxious* and *worried* are all similar in intensity.

注释: 英语中 nervous, anxious 和 worried 三个形容词的含义和用法有所不同。
碰到一些事情，譬如你要去演说，你可能在事前或在演说中会显得"紧张"，也就是显得 nervous。有些人凡事都会"紧张"，可以说有 nervous 的个性。anxious 往往用来形容"忧虑"某事可能会发生；worried 也形容"忧虑"，但"忧虑"的程度比 anxious 强烈一些。

KEY

1 He's a dentist.
2 He's talking to a patient.
3 uneasy, anxious*, nervous*, worried*, frightened, terrified, petrified
   *These three are very similar.
4 open answers

EXERCISE 8

KEY

1 Because they have had a bad experience in the past.
2 c, e
3 You should brush your teeth every night before bedtime. You shouldn't eat a lot of sweets. You shouldn't eat anything straight after brushing your teeth.

TAPESCRIPT

MUBARAK SAMJI

People are, even in this day and age, frightened of going to the dentist, and it is generally because they've had a bad experience in the past. If a patient turned up who was very frightened, you would have to go through the various stages. First of all I would try doing treatment with just a local anaesthetic, but because people are frightened of needles, I would ask them to take a tranquilliser by mouth, which in most cases makes them feel a little better. But you do also get patients who are beyond that, and for them they would have to have a tranquilliser admin, . . . administered in their vein. And you unfortunately also get the final category of patient who would need to be . . . actually be put out completely, by means of a general anaesthetic.

PRESENTER: Now listen to a conversation between Mr Samji and one of his younger patients.
MR SAMJI: Do you eat a lot of sweets?
PATIENT: Only on Saturdays.
MR SAMJI: That's good, so not, not so many.
PATIENT: No.
MR SAMJI: That's ever so important. You should eat as few as possible, because it's sweets that give you holes in your teeth. All right?
PATIENT: Yes.
MR SAMJI: And how about brushing, how often do you brush your teeth?
PATIENT: In the mornings and sometimes in the evenings.
MR SAMJI: Excellent. Brushing in the evening is ever so important; you should brush every evening before bedtime. And remember not to eat anything straight after you've brushed your teeth. All right?

FOCUS ON FUNCTIONS EXERCISE 9

KEY

1 The students should tick: I'm afraid of . . . Don't worry. It'll be all right. I'm frightened of . . . Relax.
2 a) I'm afraid of . . . I'm worried about . . . I'm frightened of . . .
   b) Don't worry. Don't cry. Don't be afraid. Calm down. It'll be all right. Relax.

TAPESCRIPT

DENTIST: Right. Open your mouth. I'm just going to give you a little injection.
MAN: Er, no! No! Sorry . . . er . . . I don't like injections . . . . I'm afraid of needles.
DENTIST: Don't worry. It'll be all right. You'll hardly feel a thing. It's quite bad, that tooth. I have to fill it.
MAN: No, really . . .
DENTIST: All right. Here we go then . . .
MAN: Stop! No, sorry, I really can't. I'm frightened of drilling too. I'd better go.
DENTIST: No, sit down again. Relax. You're not the only one, you know. Why don't we just talk about it for a moment . . .

DEVELOPING VOCABULARY EXERCISE 10

KEY

2 frightened + of   terrified + of   nervous + about/of
   petrified + of   uneasy + about   worried + about
   anxious + about

SPEECH PATTERNS EXERCISE 11

KEY AND TAPESCRIPT

A: Don't worry.
B: It'll be all right.
C: Don't be afraid.
D: Don't cry.
E: Calm down.
F: Relax.

CREATIVE WRITING: Creating a mood

Focus	SKILLS
TOPICS • Sounds • Ghost stories • Fear	• Listening: sounds, a sound sequence • Reading: a literary extract • Writing: a story

📼 COMPARING CULTURES EXERCISE 1

KEY AND TAPESCRIPT

1 rumble (i)   2 smash (h)   3 growl (f)   4 crack (d)
5 cough (l)   6 rustle (g)   7 scream (a)   8 drip (j)
9 bump (e)   10 creak (b)   11 whisper (c)
12 splash (k)

EXERCISE 4

KEY

1 a dog

2 Because the dog was standing at the door, growling.

3 In the bedroom. The narrator was in bed.

4 Because they could hear a noise in the house, not far from the bedroom.

5 They listened for the sound.

6 It was still outside, there was no wind coming through the window (casement).

EXERCISE 5

KEY

1 standing upright, straight up

2 *paralysed/frozen*: you cannot move because you are so frightened; *tense*: very nervous, holding your body stiff because of your fear

3 they are all quiet noises:
a *faint* noise: a quiet, distant noise
a *muffled* noise: a noise which is not very clear
the *moaning* of the wind: a low noise, as if made by a person in pain
the *snuffling* of a dog: a sniffing noise, as if the dog is smelling something

EXERCISE 6

KEY

1 At first, and, Then, suddenly, And then

2 Examples: first, afterwards, after that, when . . . , next, finally, to begin with

3 Example: First, the dog went to the door. Then the man woke up and he sat up in bed. At first/To begin with, he couldn't hear anything, but then he heard the sound for the first time.

EXERCISE 8

SUGGESTED KEY

1 a) very big: the depths of the house
  b) probably in an old-fashioned way, and sparsely: floorboards, casements
  c) short-haired: every hair was on end
  d) open answer

2 open answers

3 open answers

📼 LISTENING EXERCISE 9

KEY (TO QUESTION 2)

1 man – got out of bed, floorboards creaked, man coughed and spoke to dog

2 man creeps to door, dog pants and growls, man opens door, door creaks

3 dog stops in corridor, growls, man walks along and opens door

4 man closes door: nothing in room

5 bump sounds again, man frightened

6 man and dog walk to room where sound comes from

7 dog sniffs and growls at door, man opens door . . .

TAPESCRIPT

Sound sequence only.

Talkback

....

Urban survival

KEY

needle and thread   paper and pencil   phone card
paper money (enough for a taxi)   coins   safety pins
tweezers   aspirin/paracetamol   lens   plasters
whistle   tiny torch   tiny scissors   blade

7 *Away from home*

....

An unusual break

Focus	**SKILLS**
TOPIC • Places to stay **GRAMMAR** • Defining relative clauses • Relative pronouns: *who, which, that, whose*	• Reading: a magazine article • Speaking: asking questions about vocabulary

语法知识:

1. 在定语从句中, that 常用来代替其他关系代词, 特别是在会话中。

Where's the girl that sells the tickets? (= ... who sells ...)

卖票的女孩子在哪儿?

在下述单词后面更常用 that:

all; every (thing); some (thing); any (thing); no (thing); none
等等。

2. 在定语从句中，如果关系代词在定语从句中作动词的宾语，关系代词常常省去。在口语中，这种情况极为常见。

I've lost the bananas I bought this morning.

我把今天早上买的香蕉丢了。

3. 介词可放在关系代词前或放在定语从句的末尾。在英语口语中通常将介词放在定语从句的末尾并省去关系代词。试比较

The people with whom he worked regarded him as their friend.

The people he worked with regarded him as their friend.

和他一起工作的人都将他当作朋友。

Relative pronouns: who, which, that, whose 关系代词

"关系代词"在定语从句中同时起两个作用。像其他代词一样，它们用作动词的主语或宾语；同时像连词一样，它们把从句与主句连接起来。试比较：

What's the name of the blonde girl? She just came in.

那位金发女郎叫什么名字？她刚进来。

What's the name of the blonde girl who just came in.

刚进来的那位金发女郎叫什么名字？

在第二个例句中，who 代替 she 成为 came 的主语，同时把两个句子连接成一句。

GETTING STARTED EXERCISE 1

KEY

1 a) guesthouse, youth hostel, hotel, B & B

   b) guesthouse, youth hostel (some), hotel

   c) youth hostel (some), caravan park, campsite, self-catering flat

   d) caravan park, campsite

2 a) hotel

   b) guesthouse, self-catering flat

   c) youth hostel, caravan park, campsite, B & B

3 open answers

EXERCISE 4

KEY

1 The wooden front door.

2 The hotel melts, and Nils holds a contest to predict when it will happen.

3 A survival certificate.

4 It's made of ice and everyone sleeps on ice beds. It has an ice theatre, a radio station and an ice bar. The rooms have no doors, furniture or heating.

EXERCISE 5

SUGGESTED KEY

1 It is a play on the words *Arctic* and *art*. The hotel is in the Arctic Circle. Each hotel is a work of art, and the owner also organises art exhibitions.

2 There are no doors, no furniture, no heating.

3 an architect/an artist/a hotel manager/an exhibition organiser

4 open answers

DISCOVERING LANGUAGE EXERCISE 6

KEY

1 a) which, that   b) who, whose, that

2 a) false   b) true

   c) true – they usually *define* a person or thing

3 that

4 The relative pronoun can be omitted when it is the object of the verb in the following relative clause, but not when it is the subject:

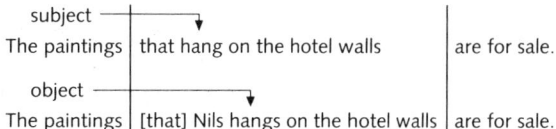

subject →		
The paintings	that hang on the hotel walls	are for sale.
object →		
The paintings	[that] Nils hangs on the hotel walls	are for sale.

EXERCISE 7

KEY

1 that/which   2 that/which   3 that/who   4 that/which – can be omitted   5 whose   6 that/which – can be omitted   7 that/which – can be omitted

Facilities

Focus	**SKILLS**
TOPIC • Hotel facilities	• Listening: interviews, a hotel dialogue
GRAMMAR • Causative: *have something done*	• Speaking: discussion, role play
FUNCTIONS • Polite requests • Asking if something is possible • Expressing satisfaction	**SPEECH PATTERNS** • Expressing satisfaction/sarcasm

语法知识：

Have something done 表示"使某事得以做成"。例：

If you don't get out of my house, I'll have you arrested.

你要是不从我家里滚开，我就叫警察把你抓起来。

GETTING STARTED EXERCISE 1

SUGGESTED KEY

1 a) fax machines, a high-quality restaurant, meeting rooms, secretarial services, mobile telephone hire, photocopiers, a laundry service, a dry-cleaning service, car hire

   b) a swimming pool, a fast food snack bar, cartoon film shows, a children's playground, a babysitting service, a laundry service

   c) a disco, tennis courts, a swimming pool, a fast food snack bar, (cartoon film shows), horse-riding, a sports centre, satellite television

d) a disco, tennis courts, a swimming pool, horse-riding, a room with a balcony, a room with a view, a sports centre, tour guides, car hire

e) a swimming pool, a high-quality restaurant, a room with a balcony and view, tour guides, car hire

🔊 Documentary

🔊 EXERCISES 3 AND 4

Note: The phrase *horseback riding* is used on the cassette. This is American English. The usual British English phrase is *horse riding* or just *riding*.

注释: 磁带中的 horseback riding 是美式英语。通常英式英语中用短语 horse riding 或 riding 。

KEY FOR EXERCISE 3

1 New Mexico
2 two tennis courts, a swimming pool with a sun deck, horseback riding
3 a) the horses   b) the tennis courts

TAPESCRIPT FOR EXERCISE 3

JOHN EGAN

We have two tennis courts at Rancho Encantado that are of a very good surface called plexipave. And the individual that we have taking care of them we refer to as the recreation director. We have a swimming pool, which has a sun deck around it, where the guests can sit back and just enjoy the beautiful sunshine of New Mexico. Horseback riding is a very important feature of Rancho Encantado and it's certainly one of the reasons that people come from around the country to, to stay at the ranch. The person in charge of our stable operation we refer to as the head wrangler. 'Wrangler', of course, is a term that is common in the south-west and it just means the person who takes care of the horses.

KEY FOR EXERCISE 4

1 No, most of the guests have never ridden before.
2 The beautiful scenery and the important history of the area of cowboys and Indians.

TAPESCRIPT FOR EXERCISE 4

WRANGLER

Most of our riders have never been on a horse. We're happy to take them out and give them an opportunity to see the land and the beauty of New Mexico, by horseback, not by car.

JOHN EGAN

I think the reason that people come to the south-west is not only because of the scenery itself – the vast spaces and the beautiful colours of the country – but also the important history and the legends that surround the south-west, insofar as cowboys are concerned, and Indians and so on. This is a uniquely American sort of legend, but at the same time it's something that I think people throughout the world really identify with and, and love to be a part of.

🔊 DISCOVERING LANGUAGE EXERCISE 6

KEY

Mr Johnson is a hotel guest and B is the receptionist. The polite language used by B and the general context make this clear.

1 someone else
2 someone else

3

HAVE +	OBJECT +	PAST PARTICIPLE
I'll have	the car	washed today.
I need to have	my bike	mended.
Do you want to have	your hair	cut?
Have they had	their order	taken?

4 We use this structure when someone else performs a service for us.

EXERCISE 7

KEY

	HAVE	OBJECT	PAST PARTICIPLE
1	They had	the walls and ceilings	painted.
2	They had	the carpets	cleaned.
3	They had	the windows	repaired.
4	They had	the doors	mended.
5	They had	the pictures	changed.
6	They had	smoke alarms	fitted.

FOCUS ON FUNCTIONS EXERCISE 8

Note: Students can find the replies to *Would you mind . . . ?* confusing. If you want to agree to the request, the reply is usually *No, not at all*. If you wish to refuse the request, the reply is usually *Well, yes, I would, because* . . . The opposite reply is given ('no' to agree and 'yes' to refuse) because *mind* carries a negative meaning.

Alternative presentation

KEY

1 a) Would you mind filling in this form, please?
   b) Can I have my suit cleaned, then?
   c) Excellent!
2 Would you mind + -*ing*.
3 a) Would you mind paying your bill now, please?
   b) Would you mind signing this, please?
   c) Would you mind letting me see your card, please?

d) Would you mind showing me your passport, please?

e) Would you mind completing this questionnaire, please?

📻 SPEECH PATTERNS EXERCISE 9

KEY

The sarcastic intonation carries a low fall.
The satisfied intonation carries a high fall.

1 A   2 B   3 B   4 B   5 A   6 A   7 B   8 A   9 B

SPEAKING: Telling stories

FOCUS	SKILLS
TOPIC • Experiences abroad	• Reading: a literary text • Listening: an anecdote • Speaking: telling, and retelling stories

READING: A LITERARY EXTRACT EXERCISE 2

Background note

Gerald Durrell is an English writer of travel books and books about the natural world. He is also known for his work as the Director of Jersey Zoo.

背景知识注释：

杰罗特·杜雷尔是一位英国作家。他专写旅游作品和有关自然世界的书籍。同时，因担任泽西动物园主任工作出色而闻名。

KEY

The order is: C, B, D, A

EXERCISE 3

KEY

1 true   2 false – He's a young man.   3 true
4 false – It hit Gerald's mother.
5 false – It hid under Leslie's plate.

EXERCISE 4

KEY

1 Noise: uttered, barking, thumped
2 Movement: scattering, sped, quivering, leapt, flicked, peered, hurled, drenched, swarm
3 a) leap   b) quiver   c) hurl   d) peer
   e) flick   f) drench

📻 LISTENING EXERCISE 7

KEY

1 It was a camping safari.
2 She stopped each night and put up her tent.

3 Because the coach broke down.
4 There was an infestation of mice – there were mice everywhere.
5 When she opened her suitcase, about six mice leapt out.

📻 EXERCISE 8

KEY

1 It's less complicated with fewer different verbs and adjectives. It's also more informal, e.g. *gear* instead of *equipment*, *stuck there* instead of *unable to travel further*. She also uses *you know* and *sort of*, which are features of spoken colloquial English.
2 With simple linkers such as *and*, *so*, *but*.
3 There is repetition to create dramatic effect and she uses her voice (intonation) to show her feelings, e.g. *There were mice everywhere, everywhere you looked*. She doesn't speak in complete sentences and doesn't always finish them.

Progress check: Units 6–7

Grammar and functions

EXERCISE 1

1 What would you do if you failed your exams?
2 When will you/we leave if there are no seats on the 14th?
3 How will he travel if the coaches are full?
4 Where would you shop if there weren't any local shops?
5 How would you feel if she needed a major operation?

EXERCISE 2

1 which – blank   2 which   3 who   4 who – blank
5 whose   6 which   7 which – blank

EXERCISE 3

1 She had the letters typed.
2 He has his shopping done.
3 I'll have the fence painted.
4 They're having/going to have some business cards printed.
5 We're going to have our picture taken.

EXERCISE 4 (SUGGESTED ANSWERS)

1 Students shouldn't study all day. They ought to enjoy themselves.
2 People oughtn't to try to change their looks. They should accept them.
3 We shouldn't have children at twenty. We ought to wait until we are thirty.

4 You should talk about your fears. You oughtn't to keep quiet.

5 We shouldn't stay at the ice hotel. We ought to go somewhere hot.

EXERCISE 5

1 Can I have my suit cleaned?

2 Calm down!

3 Would you mind closing the door, please?

4 I'm very nervous about speaking in public.

Vocabulary

EXERCISE 6

1 overconfident   2 acceptable   3 predictable
4 oversleep

8 *Paths to success*

Job options

Focus	
TOPICS	**FUNCTIONS**
• Jobs	• Talking about ambitions
• Qualifications	**SKILLS**
• School subjects	• Reading: case studies
GRAMMAR	
• Present perfect simple	
• Adverbs: *for* and *since*	

语法知识:

现在完成时常常用来表示已经完成的动作和事情。通常是在过去的事情对现在有一些重要意义时使用。

I've been all over Africa. (= I know Africa well.)

我已经走遍了非洲。(＝我很熟悉非洲。)

现在完成时也用来表示过去的而不是最近的动作,但是这些动作现在仍然是我们经历和知识的一部分。

I've travelled a lot in America. (= I know America.)

我在美国旅游了许多地方。(＝我了解美国。)

Adverbs: for and since

如果 for 表示的一段时间一直持续到现在为止,就要用现在完成时。

I've known her for a long time.

我认识她已经好长时间了。

要表示持续至说话时刻的动作或情况开始于什么时候,要用 since。

I've been here since three o'clock, but nobody's come yet.

我从三点起一直在这儿,但是没有人来过。

当 for 和 since 都用在现在完成时句子中时,这两个词很容易搞混。for 表示什么事情延续了多长时间;since 则表示这件事是从什么时候开始的。例如: for three weeks; since Monday

GETTING STARTED EXERCISE 1

KEY (FOR UK)

a) dentist, engineer, teacher, journalist, librarian, pharmacist

b) carpenter, chef, nurse, plumber, firefighter, hotel receptionist, window dresser, lorry driver, police officer

c) cleaner, postman, waiter

Note: It is becoming increasingly common to find an alternative for -*man*/*woman*, e.g. *firefighter* rather than *fireman*, *police officer* rather than *policeman*/*woman*. It is still common to use -*man*/*woman* with jobs such as *postman*/*milkman*.

READING EXERCISE 2

KEY

A is putting clothes on dummies in a shop window.

B is making something out of wood.

C is giving an injection to a horse.

EXERCISE 3

KEY

1 C   2 A   3 B

EXERCISES 4 AND 5

KEY FOR EXERCISE 4

a) Diana, Paul and Liz   b) Liz   c) Diana   d) Paul

KEY FOR EXERCISE 5

a) 'A' levels   b) degree   c) diploma, certificate

EXERCISE 6

SUGGESTED KEY

Diana: 'chemistry, bi'ology

Paul: art, craft, de'sign and tech'nology

Liz: ,mathe'matics, craft, de'sign and tech'nology

COMPARING CULTURES EXERCISE 7
Background note

Advanced level examinations ('A' levels) are the entrance qualification for universities (and higher education in general). In the UK **higher education** refers to universities and other institutions offering academic courses, which usually lead to **degrees. Further education** refers to colleges offering more vocational qualifications.

背景知识注释:

高级水平考试("A" levels) 是大学的入学资格考试。在英国,高等教育指开设专业课程并有权授予学位的大学和学术机构。继续教育指提供更多职业能力的学院。

🔊 FOCUS ON FUNCTIONS EXERCISE 8

KEY

1 1 window dresser   2 vet   3 carpenter
2 a) My ambition is to become a manager.
   b) I've always wanted to work with animals.
   c) In the future, I'd like to work for myself.

TAPESCRIPT

One.
You've got to make them look as beautiful as possible because that's your job. You're there to make the customer think: I want to look like that – exactly the same as models on the catwalks in Milan or Paris. My ambition is to become a manager so that I can decide on the themes for displays instead of just dressing the mannequins. It *is* possible to get a job doing this kind of work without qualifications, but lately it's become difficult. I'm glad I've got a proper qualification.

Two.
I prefer dealing with domestic pets because I like these smaller animals. I think the problems are more interesting; er, I like talking to all the different people who bring them in. I don't have any other ambitions, really. I've always wanted to work with animals and now I do!

Three.
You have to provide all your own tools. Saws are quite expensive, and you have to replace them every couple of months, because they wear out. The most irritating thing about the job is when I get stuff in my hair that I have to cut out, and of course my hands and clothes are always dirty. In the future I'd like to work for myself . . . to be self-employed and run my own carpentry business.

DISCOVERING LANGUAGE EXERCISE 10

KEY

Examples from the texts:
Diana Stapleton, 28, has been a veterinary surgeon for three years. (use c)
She has always loved animals. (use c)
Since then she has obtained a basic skills certificate. (use b)

1 Examples: *She hasn't been to America. What have you done in your life?*
2 a) false – past simple   b) true   c) true
   d) false – past simple
3 *For* and *since* are used to describe actions that began in the past and continue up to the present time. *For* + a period of time, e.g. a year, three seconds; *since* + exact time, e.g. yesterday, 1976.

EXERCISE 11

KEY

1 have you been   2 For   3 Has the company employed   4 has not   5 Have you taken
6 finished   7 Did you enjoy   8 loved
9 haven't used

Extra practice
Write these sentence stems on the board:
_____ *recently.*
I have never _____ .
I once _____ .
_____ *for the past three years.*
_____ *for a month.*
I have always _____ .

Preparing to work

Focus	
TOPIC • Arts and crafts GRAMMAR • Present perfect progressive • Linkers	SKILLS • Listening: an interview • Reading: an article • Writing: a brief biography

语法知识:
现在完成进行时可以特别用来表示比较短暂的动作和情况。
I've been living in this flat for the last month.
一个月以来，我一直住在这间套间里。
My parents have lived in London all their lives.
我父母一生都住在伦敦。
现在完成进行时强调动作的连续性。
Sorry about the mess — I've been painting the house.
非常乱，很抱歉 — 我一直在油漆房子。

🎬 Documentary

LISTENING EXERCISE 1
Note: *Bunnell* is pronounced /bəˈnel/.

🔊 EXERCISE 2

Background note
The Royal College of Art (RCA) is a leading art college in London and is well-known for its annual show of students' work.

背景知识注释:
皇家艺术学院 (RCA) 是位于伦敦的一所著名艺术院校。它因一年一度的学生作品展而闻名。

SUGGESTED KEY FOR EXERCISE 1

1 She's an art student.   2 a sculpture
3 'A' levels in arts subjects

KEY FOR EXERCISE 2

1 at the Royal College of Art
2 2 years
3 almost 2 years ago
4 almost 2 years
5 A show of the students' work. They also talk about their work for an hour.
6 at the beginning of September

TAPESCRIPT

KATIE BUNNELL

I'm studying for an MA in ceramics at the Royal College of Art. I've been studying for nearly two years now – this is my final year. The first year we do quite a lot of project work, and then throughout the summer holidays after the first year we have to write a thesis. And then in the second year we have a final exam, where we have to put up a show and talk about our work for an hour. And then we have a big show where the public come to see what we've made. I'm working on a project at the moment which is basically making a large piece of sculpture. I've been working on it now since the beginning of September. And I started the project by making large collages from drawings of myself. And the sculpture is made up of a pair of arms and a pair of legs; and they're very large and curvaceous; and they're going to be in very, very bright colours; and they will be set against a stripy coloured background. In the end I will present it for my MA. It'll be part of my show.

📼 EXERCISE 3

KEY

NOUN IN INTERVIEW	RELATED VERB
a thesis /ˈθiːsɪs/	write
a project /ˈprɒdʒekt/	work on, start
a sculpture /ˈskʌlptʃə/	make
a collage /kɒˈlɑːʒ/	make

EXERCISE 4

KEY

a) final   b) large and curvaceous   c) bright
d) stripy, coloured

DISCOVERING LANGUAGE EXERCISE 5

KEY

1	AUXILIARY *HAVE*	+ *BEEN*	+ *-ING* VERB FORM
2	She has You have We have They have	been	studying for nearly two years.

3 How long have you been studying?

4 She hasn't been studying at Manchester University. (She's been studying at the Royal College of Art.)

EXERCISE 6

KEY

1 How long have you been studying economics? For about a year.
2 How long have you been doing the computer course? Since January.
3 How long have you been working in the library? Since two o'clock.
4 How long have you been waiting for your interview? For half an hour.

DISCOVERING LANGUAGE EXERCISE 7

KEY

1 the present perfect simple (Sentence A)
2 There is no difference in meaning.
Note: Verbs like *work, live, drive, smoke* can be used in either tense and the choice depends on the speaker.

EXERCISE 8

KEY

1 I've been living/I've lived in this house since 1980.
2 I've been studying/I've studied English since 1990.
3 I've been writing this essay since 10.00/for five hours.
4 I've played three games of tennis this morning./I've been playing tennis since 11.00 a.m.

EXERCISE 9

KEY

1 False. He has been a student./He was a student.
2 False. He worked as a lawyer for less than a week.
3 true
4 true
5 true
6 False. He wore a suit for less than a week.
7 False. He has been making furniture in a studio for a few years.
8 true

DISCOVERING LANGUAGE EXERCISE 10

KEY

After, then, eight years later, finally, when, before, now
1 After   2 before   3 later   4 then   5 After/when

Speaking: Personal interviews

Focus	
TOPIC	**SKILLS**
• Personal information	• Reading: job advertisements
FUNCTIONS	• Listening: a job interview
• Asking for repetition	• Speaking: interviews
• Asking for clarification	**SPEECH PATTERNS**
• Correcting yourself	• Using intonation to introduce a new topic
• Showing interest	

READING EXERCISE 1
Background notes

1 **Hampstead** is an area in London, about five miles north-west of the centre. It is an expensive place to live and is considered to be a desirable area.

2 **Au pair** is pronounced /əʊ'peə/. They are usually young women who go abroad and live with a family, normally in order to learn their country's language, in return for doing light work in the house or looking after children.

背景知识注释:

1. 汉普斯特位于伦敦中心区西北五英里左右的地方。它被看作是一个理想的居住区,消费水平很高。

2. Au pair 读作 /əʊ'peə/。通常指去国外,与外国人家庭一起生活的年轻女性。她们一般是为了学习这一国家的语言。同时她们帮助主人做一些简单家务或照看孩子。

KEY

1 a live-in au pair
2 Hampstead, London
3 between 18 and 25
4 No, but they must speak good English.
5 £100 per week
6 Their own room with a television. Their fare home will be paid at the end of the contract. They may have use of a car.

🔊 EXERCISE 3

KEY

3 family   4 school/education   5 work experience
6 interests   7 experience with children

🔊 EXERCISE 4

KEY

1 He's from Helsinki, Finland.
2 He's twenty years old.
3 He came one month ago.
4 He's been visiting friends in Scotland and Wales, and looking for a job.
5 He's got two brothers and one sister.
6 His brothers are ten and eight and his sister is eighteen.
7 He's a student at Helsinki University.
8 He's studying English, French and Business Studies.
9 He's been studying there for two years.
10 He's worked in a paper factory, he's been a cleaner, he's worked in a shop and at EuroDisney.
11 Yes he has. He worked with children at EuroDisney and he's been a helper at a youth centre in Finland. He's also looked after his brothers.

TAPESCRIPT

MRS SANDERS: Hello, I'm Margaret Sanders. You must be Emil.

EMIL: Yes, Emil Aalto.

MRS S: How do you spell that?

EMIL: Emil's E-M-I-L. And my surname's Aalto . . . A-A-L-T-O.

MRS S: And where are you from, Emil?

EMIL: From Finland – Helsinki.

MRS S: Oh really? How old are you?

EMIL: Twenty.

MRS S: Sorry?

EMIL: I'm twenty years old.

MRS S: So how long have you been in Britain?

EMIL: Oh . . . about a month.

MRS S: And did you come specifically to find a job?

EMIL: No, not really. I arrived a month ago and I wanted to travel, you know . . . and improve my English . . . but then I thought it would be a good idea to try to find work.

MRS S: And what have you been doing in the last month? Travelling around?

EMIL: Er . . . yes . . . a little. I've been visiting friends in Scotland and Wales. And I've been looking for a job.

MRS S: Yes, of course. Right. Can I ask about your family? Have you got any brothers and sisters?

EMIL: Yes. I've got two brothers and one sister.

MRS S: I see. And how old are they?

EMIL: Oh, well, my brothers are quite young. One is ten and the other's eight. My sister's closer to my age . . . she's . . . er . . . eighteen now.

MRS S: So you're the oldest . . .

EMIL: Yes, of course. I don't live at home any more, but I often see them at weekends.

MRS S: Right. What about school? You went to school in Helsinki?

EMIL: Yes, that's right.

MRS S: And what were your favourite subjects?

EMIL: Oh, well ... I liked languages, geography ... and music. They were my favourites I think. And then I decided to study languages at university.

MRS S: Oh. Which university?

EMIL: In Helsinki. I started there two years ago. I'm doing English, French and Business Studies.

MRS S: I'm afraid I don't understand. You haven't finished your course yet?

EMIL: No. I decided to take a year off to improve my English before the final exams.

MRS S: Oh, I see. Good idea. OK. What about work? Have you got any work experience?

EMIL: Yes. I had a job in a paper factory for a few months after I left school. Then while I was a student I worked in the summer holidays ... first as a cleaner, then in a shop ... then I got a job at EuroDisney.

MRS S: Near Paris?

EMIL: Yes, that's right. It was good because I was able to use my English and my French.

MRS S: What was the job exactly?

EMIL: I had to dress as a bear ... and entertain children.

MRS S: A bear? That must have been fun.

EMIL: Oh yes. I really enjoyed it. But it was quite hot in the costume sometimes.

MRS S: I can imagine. OK ... Do you drive? Have you got a driving licence?

EMIL: Yes, I have. A Finnish licence and an International one.

MRS S: That's fine, then. So, what kind of interests do you have?

EMIL: Interests? Well, I like travelling ... I play a lot of sports ... and I play the piano.

MRS S: What sort of sports do you like?

EMIL: Football, tennis and swimming. I ski as well.

MRS S: Right. And what sort of music do you play on the piano?

EMIL: Oh. A lot of different types. Classical, jazz, blues ...

MRS S: Really? Are you a good pianist, then?

EMIL: Well ... not as good as I'd like to be!

MRS S: Right. The most important questions now. What experience have you had with children?

EMIL: Well, I've looked after my brothers, as babies and as young children. I've also worked with children in a youth club.

MRS S: A youth club?

EMIL: Yes. I've been working as a helper at a youth club in Helsinki since I started at the university ... as a

sort of volunteer ... with teenagers ... you know, helping them to organise things they're interested in.

MRS S: That's wonderful.

EMIL: And then, of course, I've worked with younger children as well – at EuroDisney. I haven't looked after children in a family before, but I'm sure I'll enjoy it.

MRS S: Good. Now, have you got any questions you'd like to ask?

EMIL: Er ... yes. What are the times?

MRS S: The times?

EMIL: Er ... sorry. I mean, what are the hours ... the working hours?

MRS S: Well ... they'll have to be a bit flexible, but usually while we're at work, Monday to Friday ... about eight in the morning till six in the evening.

EMIL: I see. And the pay is £100 ...

MRS S: Yes, £100 a week ... and ... er ... accommodation and food are included, of course. You'll be able to use a car during the day and sometimes in the evenings. Then, as it says in the advertisement, we'll pay your fare back to Helsinki at the end of the contract.

EMIL: Right ... could I ask a few other things?

MRS S: Of course.

EMIL: The contract is for a year ... what about holidays in that time? Will I be able to ...

🔲 SPEECH PATTERNS EXERCISE 5

KEY AND TAPESCRIPT

We generally use a falling tone when using a word to change topic.

MRS SANDERS: So, how long have you been in Britain?

MRS SANDERS: Right. Can I ask about your family?

MRS SANDERS: OK. Do you drive?

MRS SANDERS: Now, have you got any questions you'd like to ask?

🔲 EXERCISE 6

KEY AND TAPESCRIPT

A sentence ending in a low fall (e.g. a and b) can sound uninterested. A high fall or a rise at the end sounds interested and friendly.

a) A: I've got a brother who lives in New York.
   B: [uninterested] Really?

b) A: I've got a brother who lives in New York.
   B: [uninterested] Oh.

c) A I've got a brother who lives in New York.
   B: [very interested] Have you?

d) A I've got a brother who lives in New York.
   B [very interested] New York?

e) That's interesting.   f) That's terrible.   g) Do you?

h) How fantastic!   i) How nice!   j) How awful!

k) I see.   l) Oh dear.

📼 FOCUS ON FUNCTIONS EXERCISE 8

KEY

1 a, b, f

2 c, e, g

Note: It is possible to use *Sorry?* or *Pardon?* to ask for clarification but usually in conjunction with a question, e.g. *Sorry? What do you mean?*

3 d

📼 LISTENING EXERCISE 9

KEY

a) present simple   b) present simple   c) past simple

d) past simple and present perfect simple

e) present simple

TAPESCRIPT

A I'm twenty years old.

B Oh, well, my brothers are quite young. One is ten and the other's eight. My sister's closer to my age . . . she's . . . er . . . eighteen now.

C Oh, well . . . I liked languages, geography . . . and music. They were my favourites I think. And then I decided to study languages at university.

D Yes. I had a job in a paper factory for a few months after I left school. Then while I was a student I worked in the summer holidays . . . first as a cleaner, then in a shop . . . then I got a job at EuroDisney. I've looked after my brothers, as babies and as young children. I've also worked with children in a youth club.

E I like travelling . . . I play a lot of sports . . . and I play the piano.

9 *Skin deep*

First impressions

Focus

TOPIC

- People's appearances

GRAMMAR

- *Look* + adjective
- *Look like* + noun

FUNCTIONS

- Certainty and probability in the future

SKILLS

- Listening: a discussion
- Speaking: a discussion
- Writing: a short report

VOCABULARY DEVELOPMENT

- Adjectives describing people

语法知识:

Look 的一个意思与 appear 和 seem 相同。假如我们说一个人 looks tired 或 looks angry，我们的意思是他似乎"疲劳"或"生气"，他用他的表情和行为表现出来。在这个意义上，look 之后跟形容词而不是副词。类似的动词还有: be, seem appear, sound, feel, smell, taste。

You look very unhappy — what's the matter?

你看来很不高兴 — 怎么回事?

Look like + noun

在 look like 这一动词词组中，like 的用法跟介词相似: 后面跟名词、代词宾格（me, him 等）及带 -ing 的动词形式。

His sister looks like him.

他姐姐的相貌像他。

📼 EXERCISE 2

KEY

Yes, they do recognise him. The first speaker says, 'He looks awful these days' comparing his appearance to the way he used to look. Comments like 'He's his own person' and 'He's got so much money' show that they have read or heard about him.

TAPESCRIPT

JENNY: Look at him! He looks awful these days, doesn't he?

GAIL: No, he doesn't. He looks nice. I like him. He's his own person.

JENNY: Oh, come on. He looks middle-aged, and his clothes . . .

GAIL: What's wrong with them?

JENNY: Well, he looks like an old tramp . . . and he's got so much money.

DISCOVERING LANGUAGE EXERCISE 3

KEY

1 a) age – C

b) appearance – A, (B), D

c) personality – B

2 *He looks* + adjective

He looks like + noun

DEVELOPING VOCABULARY EXERCISE 4

SUGGESTED KEY

See chart above for questions 1 – 3.

4 open answers

5 He's had his ear pierced. He's started to wear glasses. He's started to dress more casually.

COMPARING CULTURES EXERCISE 5

Background note

Nigel Kennedy is a famous British classical violinist, who became well-known for his punk appearance when playing classical pieces.

背景知识注释：
奈杰尔·肯尼迪是英国著名的古典音乐小提琴演奏家，以在演奏古典曲目时的崩克打扮而出名。

LISTENING EXERCISE 6

KEY

The man has grown a beard and his hair. He's had his hair dyed and he's bought a wig. He's had his ear pierced. (He's had his teeth straightened.)
The woman has grown her nails. She's had her head shaved. She's had her arm tattooed and her nose pierced.

🔲 EXERCISE 7

KEY

1 a) Pam   b) Lucy   c) Jenny
2 a) Lucy   b) Pam   c) Jenny
3 Yes, they are very familiar with each other and use expressions used amongst friends, e.g. *You must be joking!*
4 She's probably in her late teens/early twenties: she has been dyeing her hair for a few years, but is still influenced by her parents (*Your mum'll go crazy!*).

🔲 FOCUS ON FUNCTIONS EXERCISE 8

KEY

1 *I'm certain (I will)*: I'll definitely
2 *There's a strong chance (that I will/won't)*: I probably won't, I'll probably
3 *There's a small possibility (that I will/won't)*: I think, I'm not sure, it's possible

TAPESCRIPT

Part One.
JENNY: What about you, Lucy? Will you ever dye your hair?
LUCY: Oh yes. I do dye it sometimes, and I'll definitely dye it again.
JENNY: Oh, I didn't know that.
LUCY: It was bright red a few years ago, then I was blonde for a while. Now I'm my natural colour again.
PAM: I'll probably dye my hair when I'm older – when I start going grey. I'm sure lots of middle-aged women do. Won't you, Jenny?
JENNY: Mm . . . I'm not sure. It looks so artificial. You can always tell when someone's hair isn't natural . . . no, I probably won't have mine dyed.
Part Two.
LUCY: I've decided to have a tattoo.
JENNY: Lucy! You must be joking! You'll regret it . . . And your mum'll go crazy!

LUCY: Well . . . maybe. But that's her problem. I've thought about it for a long time and now I'm sure I want one. Just a small one.
PAM: What sort of tattoo?
LUCY: I'm not sure at the moment . . . maybe a butterfly. Something colourful, anyway.
PAM: Where are you going to have it?
LUCY: On my shoulder, I think. Or on my arm.
JENNY: Would you have yourself tattooed, Pam?
PAM: Well, . . . it's possible. They can look quite good but they're so difficult to remove after you've got one.
JENNY: Well, there's one thing I'm certain about . . . you're both mad!

A professional interest

Focus	
TOPIC	**FUNCTIONS**
• Models and model agencies	• Guessing • Making deductions
GRAMMAR	**SKILLS**
• *Must, can't, might, could* + *be*	• Listening: a monologue, sound sequences
	SPEECH PATTERNS
	• Running words together

语法知识：
在情态动词中，must 用于讲述某种情势是确定的；might, could 表示可能，而 can't 表示不可能
You must be tired.
你一定累了。
That can't be John.
那人不可能是约翰。
Things might be worse than they seem.
情况或许比表面看上去更糟。
We could all be rich one day.
有一天或许我们大家都会发财。

🔲 Documentary

LISTENING EXERCISE 1

KEY

1 It is given to prospective models by people who work for the Elite agency.
2 They are models.
3 open answers

EXERCISE 2

KEY

1 f   2 g   3 b   4 e   5 d   6 a   7 h   8 c

📼 EXERCISE 3

KEY

1 They use a scouting programme. Each booker carries cards to give to people who may be suitable models. These contain information about present models and show that the agency is bona fide. The agency operates an open-door policy and all potential models have an expert assessment to see if they are suitable.

2 symmetrical features, height of about 5 ft 8 inches, nice eyes, nice mouth, a great personality and good professional attitude

TAPESCRIPT FOR EXERCISES 3 AND 5

CHRIS OWEN

We have a, quite a good scouting programme. A lot of the bookers will constantly, when they go home in the tube or, or if they're going away on holiday somewhere – you know, it's part of their job – and they look for a look that they think might or might not be worthwhile. This is a card that all the bookers take around with them – they keep it in their handbags. So, if they see a girl on the street that they think is particularly interesting, you know, they can take it out of their pocket or their purse (whatever it is). And it gives the would-be model a feeling of reasonable security that it is a bona fide agency, with pictures of girls on and a telephone number that they can call afterwards. And that's just one of the little things that we use to help us scout.

And we have an open-door policy: I mean, if anyone wants to come in, they come in; we can assess them; they don't need an expensive portfolio; they don't need anything really but themselves, and perhaps their mother or father to reassure them. And they can come in and they'll have an expert assessment, basically. I can show you a book of a very new girl to give you an idea of the sort of pictures that come just from testing; and then gradually she'll start doing shots for *Looks* magazine, for example, *19*; young girl magazines. We look for a girl around 5ft 8, who's got symmetrical features – you know, nice eyes, nice mouth; and girls with great personalities as well; and, you know, good professional attitudes.

📼 EXERCISE 4

KEY

Susie: b, c   Cathy: a, c, d   Joanna: b, e, f
The picture on the right shows Cathy.

TAPESCRIPT

This girl here, Cathy, is, is half-Chilean and half-English. And she was scouted in a, a shoe shop in, in the Kings Road – she was working there. And she has a very exotic Latin look. She's a very beautiful girl and again has a good personality. She's, you know . . . she's got a dark coffee skin, dark hair, and she looks great with her hair up – nice neck. She has a good profile, and photographers will like that. I mean, it creates a nice image, and I think that, you know, she is going to do very, very well, this girl, very nice.

This is another girl here, Joanna Rhodes, who's, was discovered in a model competition with *Company* magazine, and she's one of the top English models. I think a lot of English models don't necessarily look English. For example, Susie Big; she might be Spanish, she might be Italian, she might be, er, South American, you know, in fact, she's very English.

EXERCISE 7

KEY

Possible: might be, could be
Certain (+): must be
Certain (-): can't be

EXERCISE 8

KEY

1 can't be   2 might be, could be   3 must be
4 could be, might be

📼 SPEECH PATTERNS EXERCISE 9

KEY

a) *t* is not pronounced /ˈmʌsbɪ/
b) *d* is not pronounced /ˈkʊbɪ/
c) *t* is not pronounced /ˈmaɪbɪ/
d) *t* is not pronounced and *n* sounds more like an *m* /ˈkɑːmbɪ/

📼 EXERCISE 10

KEY

1 underground station   2 café   3 swimming pool

WRITING: Describing appearances

Focus	SKILLS
TOPICS • Unusual models • Ordinary people • Ideas of beauty	• Reading: an article • Writing: descriptions of people VOCABULARY DEVELOPMENT • Compound adjectives and prepositional phrases

READING EXERCISE 1

KEY

1 D, E and F are professional models.
2–4 open answers

EXERCISE 3

KEY

1 False. Some models can be ordinary looking.
2 False. In the world of fashion modelling, ordinary people look out of place.
3 False. They get more work than ugly women.
4 True.
5 False. There is an agency offering models that look like real people.

DEVELOPING VOCABULARY EXERCISE 5

KEY

1

	ONE-WORD ADJECTIVES	COMPOUND ADJECTIVES	PREPOSITIONAL PHRASES
Real people & UGLY models	ordinary, ugly, short, fat, tall, thin, dull	unusual-looking strange-looking	
Fashion models	slim, beautiful	sun-bleached well-oiled muscle-bound	with perfect skin with perfect teeth

2

	ORDINARY PEOPLE	FASHION MODELS
One-word adjectives	plump, pretty, skinny	elegant, stunning
Compound adjectives: a) adjective + noun + -ed	old-fashioned greasy-haired	lightly-tanned smooth-skinned golden-haired
b) adjective + present participle	dull-looking easy-going	good-looking fun-loving
c) adverb + past participle	badly-groomed	well-dressed
Prepositional phrases	with greasy hair with crooked teeth with large ears	with manicured nails with translucent skin

Progress check: Units 8–9

GRAMMAR AND FUNCTIONS

EXERCISE 1

1 looks   2 looks like   3 looks   4 looks like
5 looks   6 don't look

EXERCISE 2

1 must be   2 could be, might be
3 could be, might be   4 can't be

EXERCISE 3

1 has been living   2 for   3 has been trying
4 has already had   5 have been   6 has finished   7 for
8 have been doing   9 since   10 haven't achieved
11 Since   12 has been learning

EXERCISE 4

1 She'd like to work in films.
2 She's always wanted to work in films.
3 Her ambition is to work in films.

EXERCISE 5

1 No, I'm not sure. No, it's not definite.
2 Yes, almost certainly. Yes, probably.
3 Yes, maybe. Yes, it's possible.

Vocabulary

EXERCISE 6

Suggested answers:
a) clean, interesting, serious, smart, fit, relaxed, sensitive
b) dirty, scruffy, immature, crazy, aggressive, forgetful, rebellious, untidy
c) boring, clean-shaven, bald-headed, unshaven, good-looking, old-fashioned, fun-loving, rich

10 Showtime

Story-telling

Focus	
TOPIC • Puppets **GRAMMAR** • *Be able* + *to* + infinitive • *Manage* + *to* + infinitive **FUNCTIONS** • Talking about achievement	**SKILLS** • Reading: an informative text • Speaking: considering possibilities **VOCABULARY DEVELOPMENT** • Related words

语法知识:
Be able to 加动词不定式表示某人有能力做某事。
He is able to repair his own car.
他能修理自己的车。
Manage to 表示成功地做完了某事。例如:
I managed to finish the paper before Monday.
我终于在周一前写完了论文。

READING EXERCISES 2 AND 3

Note: *Punch and Judy* is pronounced /ˈpʌntʃənˈdʒuːdɪ/, *Tchantchès* /ˈtʃænˈtʃez/ and *wayang kulit* /ˈwaɪˈjæŋkʊlɪt/.

	COUNTRY	NAMES OF PUPPETS	MADE OF
Picture A	Britain	Punch, Judy, Toby the dog, etc.	wood/cloth
Picture B	Indonesia (Java)	——	leather
Picture C	Belgium	Tchantchès	wood

KEY FOR EXERCISES 1 AND 2

See grid above.

KEY FOR EXERCISE 3

1 a) puppets from Belgium
   b) puppets from Britain and Java
2 Punch and Judy shows are about 30 minutes. *Wayang kulit* shows can last all night.
3 Tchantchès and Mr Punch
4 open answers

EXERCISE 4

KEY

1 ˈlazy   2 ˈgreedy   3 ˈsimple   4 warmˈhearted
5 ˈquarrelsome /ˈkwɒrəlsəm/   6 ˈtender

DEVELOPING VOCABULARY EXERCISE 5

KEY

NOUN	ADJECTIVE
traˈdition	traˈditional
wood	ˈwooden
ˈbasis	ˈbasic
ˈcomedy	ˈcomic
ˈquarrel	ˈquarrelsome
ˈhistory	hiˈstorical
reˈligion	reˈligous

VERB	NOUN
enterˈtain	enterˈtainment
ˈvary	variˈation
ˈargue	ˈargument
perˈform	perˈformance

DISCOVERING LANGUAGE EXERCISE 6

KEY

1 can
2 *to* + infinitive (*to raise/ to see/ to tell*)
3 . . . the puppeteers *are able to raise* the arms and legs . . .
4 *Will* people *be able to see* the traditional characters for many more years? People *won't be able to see* the traditional characters for many more years.
5 *Does* he *manage to tell* a long and complex story without a script? He *doesn't manage to tell* a long and complex story without a script.

6 She *managed to solve* the problem.
   Manage + *to* + infinitive is used to talk about possibility or ability when it is difficult to do something.

EXERCISE 7

KEY

1 managed, were able, manages
2 Will you be able, won't be able
3 managed, weren't able

Stages

Focus	
TOPIC • Theatre	• Expressing possibility • Giving permission • Talking about prohibition • Making deductions
GRAMMAR • *Can* and *could*	
FUNCTIONS • Making requests • Expressing ability	SKILLS • Listening: an interview • Reading: a literary extract • Speaking: role play

语法知识：

can 和 could 常用来提议为某人做某事或请求别人做某事。在语气上，could 没有 can 那么肯定，使提议或请求听起来更加客气。

GETTING STARTED EXERCISE 1

KEY

1 open answers
2 stage – 6   curtain – 2   stalls – 5   box – 4
   dress circle – 3   upper circle – 1
3 Open answers, but should include phrases like *Can I help you? I'd like two tickets in the stalls . . . Which performance . . . ?*

▣ Documentary

▭ LISTENING EXERCISE 2

KEY

1 It's the oldest theatre in London.
2 1818
3 Queen Victoria
4 48 (major theatres)
5 television; spectacle – outdoor concerts, rock and pop music
6 yes

TAPESCRIPT

ANDREW LEIGH
Part One.
The Old Vic is a theatre, and it's the oldest theatre in

London. It was built in 1818. It was called the Old Vic because it was originally the Royal Victoria Theatre – after Queen Victoria, but popularly became known as the Vic, and then the Old Vic. Like most theatres, people sit in rows – in straight rows – facing the stage. And what we have here is the theatre divided into three levels: the stalls, the dress circle, and the upper circle.

Part Two.

There are about 48 major theatres in London with a broad mixture of presentations. Those theatres between them present anything from large-scale musicals – like *Les Miserables* or *Phantom of the Opera* or *Miss Saigon* – to small-scale plays, often new works by new writers. I think theatre styles change. We are, we have been, very influenced by television, for example, in the kind of drama we present and the way we present it. But in recent years we've become more influenced by spectacle, like outdoor concerts, for example, and the world of rock and pop music – all of this has influenced the way we do theatre; so there's now, at this time, an emphasis on spectacular theatre on a large scale. But, given those influences, I believe that the theatre will always – live theatre – will always have a place in ... as part of our British entertainment scene.

🔊 EXERCISE 3

KEY

2 this evening   3 two   4 the stalls   5 K14 and K15
6 £30   7 7.45

🔊 EXERCISE 4

KEY

a) Can I help?
b) Do you have any tickets available for tonight's performance?
c) How many was it for?
d) Yes, that's fine, thanks.
e) It starts 7.45 this evening.
Suggestions for other expressions:
a) Are you being served? Can I help you?
b) I'd like two tickets for ...? Are there any tickets left for ...? Have you got any tickets for ...?
c) How many (tickets) would you like?
d) They're fine, thanks. Yes, I'll take those, thanks.
e) The performance is at 7.45. The curtain rises at 7.45.

TAPESCRIPT

WOMAN: Hello.
TICKET CLERK: Hello, Can I help?
WOMAN: Yes, do you have any tickets available for tonight's performance?
CLERK: Yes, we have: stalls or dress circle, £30 each. How many was it for?

WOMAN: Just two, thanks.
CLERK: Two tickets. I can do two in the stalls: row K, 14 and 15 – that's eight rows back.
WOMAN: Yes, that's fine, thanks, yeah.
CLERK: Two £30 tickets ... that's £60. Thank you.
WOMAN: Thank you.
CLERK: It starts 7.45 this evening.
WOMAN: OK then. Thanks very much. Bye.

🔊 EXERCISE 5

KEY

The ticket clerk is impolite. She is irritated by the interruption. The man is also rude. Both use very direct language.

TAPESCRIPT

TICKET CLERK: Yes?
MAN: I want some tickets for tonight.
CLERK: How many? They're £30 each ...
MAN: Two.
CLERK: That's £60.
MAN: Here.
CLERK: Tickets. 7.45. Don't be late.

FOCUS ON FUNCTIONS EXERCISES 7 AND 8

KEY FOR EXERCISE 7

1 a) ability   b) possibility   c) prohibition   d) request
  e) permission   f) deduction
2 All the sentences have the same function. Sentence D is more polite with *could*.
  a, b, c, e are now in the past:
  a) My brother, the actor, could learn lines quickly *when he was younger.*
  b) I'm sorry, I couldn't meet you after work *yesterday.*
  c) She's under eighteen, so she couldn't see the film *at the weekend.*
  e) *In the 1980s,* at eighteen we could see any film at the cinema.

KEY FOR EXERCISE 8

Can is not used to describe a future ability.
Could is not used to describe a single opportunity/possibility in the past. However, *couldn't* is used to describe a single possibility in the past.

EXERCISE 9

KEY

1 possibility – couldn't/weren't able to get
2 ability – could/were able to speak
3 permission – can
4 ability – will be able to
5 possibility – couldn't/wasn't able to

6 possibility – was able to get
7 request – can/could
8 possibility – couldn't/weren't able to/were able to get

READING: A LITERARY EXTRACT EXERCISE 10

KEY

1 Shakespeare is comparing a person's life to a play performed on a stage.
2 *Man* and *his* refer to a human being.
3 And all the men and women are just actors:
They are all born and they all die;
And one person in their lifetime has many different roles;
Their life consists of seven periods.
Note: *Their* is used here to replace *his*. It is becoming increasingly common to avoid the use of *his/her* if possible and to use the impersonal pronoun *their* instead.
4 open answers
Shakespeare's seven stages are the infant; the school child; the lover; the soldier; the lawyer; the small, old man; the very old man who has lost everything (e.g. his teeth, his sight, his taste, his mind, his hearing, etc.).

WRITING: Giving opinions

Focus

TOPICS
• Circuses
• Animal performers

FUNCTIONS
• Giving opinions

SKILLS
• Reading: an article, a letter
• Listening: monologues
• Speaking: discussion
• Writing: a letter

VOCABULARY DEVELOPMENT
• Negative prefixes

SPEECH PATTERNS
• Using stress to show disagreement

GETTING STARTED EXERCISE 1

KEY

1 a) ringmasters   b) clowns   c) acrobats, lion tamers, fire eaters, human cannonballs
2 Open answers, but likely to be horses, big cats and elephants, and perhaps sea animals, such as seals.

EXERCISE 3

KEY

1 Because he likes them.
2 Yes, it is.

3 They live in small cages – the conditions are not natural for them.
4 They live longer in a circus.
5 Because the animals wouldn't normally do such circus acts in the wild.

EXERCISE 4

KEY

1 a) lion, tiger, panther, leopard /ˈlepəd/, puma /ˈpjuːmə/
   b) seals, dolphins, sharks
   c) wounds /wuːnz/   d) a cage
2 a) to tame   b) mauled /mɔːld/   c) tickled

DEVELOPING VOCABULARY EXERCISE 5

Note: The six prefixes have the meaning of *not*: *ir-* is used with adjectives which begin with *r*, e.g. *irregular*, *il-* is used with adjectives that begin with *l*, e.g. *illegal* and *im-* is used with some adjectives that begin with *p-*, e.g. *impossible* and with *m*, e.g. *immoral*.

KEY

UN-	unnatural unusual unpleasant unkind	DIS-	dissatisfied dishonest
IN-	insensitive incorrect	IM-	impractical impossible impatient immoral
IR-	irregular irrational	IL-	illegal illogical

READING EXERCISE 6

KEY

1 Because a circus is visiting the town and Mr Andrews objects to the use of animals in circuses.
2 a) paragraph 4: Why don't circuses provide . . .? Let's leave
   b) paragraph 2: I feel that, in my opinion, I believe, I am sure

📼 LISTENING EXERCISE 7

KEY

1 Speaker 3.
2 Speaker 1 thinks that only domestic animals should be used.
Speaker 2 thinks that you can't have a circus without animals and that they like performing. They should be well looked after.

3 Speaker 1: Only domesticated animals should be used: people are now used to seeing wild animals on television.Wild animals shouldn't be taken from their natural world. They shouldn't be kept in cages, as it is unkind.

Speaker 2: You can't have a circus without animals: the animals should be well cared for with big enough cages and plenty of exercise. Animals like to play. Animals have been performing in circuses for hundreds of years.

Speaker 3: Animals shouldn't be used in circuses at all: you shouldn't keep them in cages. It's wrong to make them perform acts which they don't do naturally in the wild. People don't like to see animals humiliated.

TAPESCRIPT

One.

Circuses are traditional. They've been around for hundreds of years and I'd hate to see them disappear. But they have to change with the times. People used to like seeing wild animals there . . . perhaps because they'd never seen them before . . . I mean before television, and before foreign travel was popular. But some of those animals – the tigers and the lions . . . and even the zebra . . . their numbers are so small in the natural world that we shouldn't take them away from it and put them in circuses for our pleasure. Of course, we're also more aware of cruelty these days, and keeping large animals in cages all their lives is unkind. But I'm not worried about domestic animals in circuses – certainly dogs and horses are OK. They don't need to be kept in cages all the time and if they're not asked to do dangerous tricks, I don't see the problem. We could make a law banning wild animals from circuses, but let's keep some animals.

Two.

I really don't agree. As far as *I'm* concerned, this protest is all a lot of rubbish! I loved the circus when I was younger . . . and animals are part of the circus tradition. You just can't have a circus without animals – it wouldn't be a circus. By all means, let's make sure that the animals have cages that are big enough . . . and proper exercise . . . and that they're well looked after. I don't see what the problem is about performing. I mean, all animals like to play, and when you're at the circus you can see they enjoy it . . . or at least they don't find it unpleasant. Why don't people just leave circuses alone and let them go on as they have been for the last two hundred years?

Three.

I think it's awful . . . the whole thing. In *my* opinion, any use of animals that involves keeping them in cages is wrong. And then to get them to do things that they never do naturally . . . like lions standing up on their back legs . . . or seals clapping their flippers together . . . I find the whole thing disgusting. I think all circuses with performing animals – even dogs . . . or birds – should be banned. I don't believe most people like to see animals humiliated. It's unnatural and completely inexcusable. Why don't we just leave them alone to get on with their lives? Surely circuses can come up with entertaining shows without this kind of exploitation?

SPEECH PATTERNS EXERCISE 8

KEY AND TAPESCRIPT

I think . . . In *my* opinion . . . As far as *I'm* concerned . . .

Talkback

Picture clues

KEY

There are four people in a team. One person draws a phrase (they have been given) and the other three have to guess the phrase, title of a book, etc. The person drawing the picture can only say *yes* or *no*. When they have guessed correctly, the person who gave the correct answer runs to get the next phrase, and then draws it. The team which wins is the one which guesses all the phrases first.

TAPESCRIPT

PRESENTER: Listen to people playing a game. Picture A.

WOMAN A: What's that? It looks like a potato or something . . . is it a vegetable?

MAN A: No . . . it's . . .

WOMAN A: Hey! Hey! You mustn't say any more!

WOMAN B: Oh look . . . it's got eyes . . . yes, they're definitely eyes. It must be someone's face.

MAN A: Yes.

MAN B: I know . . . it's a famous person! A man . . .

MAN A: No.

PRESENTER: Listen again. Picture B.

WOMAN B: What are those things coming out of his eyes? Is he crying?

MAN A: No . . . oh . . .

WOMAN B: They could be glasses . . .

WOMAN A: No, no, no . . . they're arrows . . .

MAN A: Yes . . . come on . . . oh, sorry.

WOMAN A: He's, he's looking in a particular direction . . .

MAN A: Yes . . . yes . . .

MAN B: Looking . . . is that one of the words?

MAN A: Er . . . yes . . .

PRESENTER: Listen again.Picture C.

WOMAN B: What's that? It looks like a picture . . .

MAN A: No.

WOMAN A: It's a book ... Looking, ... looking ... book. I know. I've got it.
MAN B: What is it then?
WOMAN A: It's *Look Ahead*!
MAN A: Yes, that's it! Well done. Your turn. Quick! Go and get the next one!
WOMAN A: Oh, OK.

KEY

D – radio station   E – fashion designer

11 *Looking forward*

∙∙

Energy

> **Focus**
>
> **TOPICS**
> • Energy sources
> • Life in the future
>
> **GRAMMAR**
> • *Will* + (adverb) + infinitive
> • *Going to* + infinitive
> • *May/might* + infinitive
>
> **FUNCTIONS**
> • Making decisions, promises, predictions
> • Expressing plans and intentions
> • Talking about possibilities
>
> **SKILLS**
> • Speaking: negotiating
> • Listening: an interview, brief comments
> • Writing: sentences justifying a decision
> • Reading: a literary extract

语法知识:

英语中有几种不同的动词形式可以用来表示将来时。常见的结构有:(1) shall/will 表示将来;(2) going to 结构。

I'll see you next week.
下周再见。

Who's going to look after the baby tomorrow?
明天谁照看孩子?

但是,我们必须注意以下几点:

(a) 当我们谈及已经决定的事情时,就用 going to 结构,一般不用 shall/will 形式。

She's going to have a baby in June.
她 6 月要生孩子。

(b) 当我们在决定做某事的当时谈到此事,要用 will (一般要用 will 的缩略式 'll)。

I'm tired. I think I'll go to bed.
我累了,我想去睡觉了。

(c) 当我们提及条件时 (即如果一事发生,另一事也就将发生),一般要用 shall/will 形式,不用 going to 结构。

If I give you money, you'll only spend it on drink.
如果我给钱,你只会买酒喝。

(d) 有时候我们说要发生什么事了,因为我们可以看出会发生

这样的事 看到了要发生这种事的迹象。在这种情况下,我们一般都用 going to 结构。

Look — it's going to rain.
瞧,要下雨了。

May/might + infinitive 的最普通的用法是谈论可能性,或要求(和给予)许可。

You know, I think it might rain.
你知道,我认为可能会下雨。

"May I have some more wine?" — "Yes, of course you may."
"我可以再喝点酒吗?" — "可以,当然可以。"

May 和 might 常用来谈论一种可能性:某事可能将要发生或某事可能正在发生。要注意的是 might 不是 may 的过去式,它所表示的可能性比 may 所表示的可能性小一些。

GETTING STARTED EXERCISE 1

KEY

1 solar power – A   coal – G   gas – H
wind power – C   nuclear energy – B   wood – F
petrol – D   hydroelectric power – E

2 and 3

FOSSIL FUELS	ALTERNATIVE ENERGY SOURCES	RENEWABLE	NON-RENEWABLE
coal	solar power	solar power	coal
gas	wind power	wind power	gas
wood	nuclear energy	wave power	wood
petrol	wave power	nuclear energy	petrol

🎬 Documentary

📼 EXERCISE 3

KEY FOR EXERCISE 2

1 It's a plant for producing electricity.
2 Solar power: the manufacturing of electricity by using sunlight.

KEY FOR EXERCISE 3

1 He provides electricity for the city of Austin in Texas.
2 Photo = light, voltaic = of electricity. It means using sunlight to make electricity.
3 100 – 150 homes
4 We will use fewer non-renewable forms of energy (coal, gas) and more renewable ones e.g. wind power.

TAPESCRIPT

KEVIN: People who work in the energy industry – making electricity – have to look into the future to predict our needs. Especially people who work in 'alternative energy' – people who look for new ways of making electricity, such as photovoltaics. The man to explain photovoltaics and look into the future is John Hoffner. His job is to provide electricity for the city of Austin in Texas. So what exactly does 'photovoltaics' mean?

JOHN: Photovoltaics simply means converting light into electricity. 'Photo' means 'light' and 'voltaic' means 'of electricity'. So the simple definition is that it converts sunlight to electricity. We in Austin have the second largest photovoltaic plant in the United States. That plant is about 300 kilowatts in size; that's enough electricity for about 100 to 150 homes in Austin.

KEVIN: What will our sources of energy be fifty years from now?

JOHN: Fifty years from now we will see a completely different energy picture: a lot less reliance on traditional forms of electricity such as coal and natural gas and oil, and a lot more reliance on renewable energies such as photovoltaics, wind and hydro power.

EXERCISE 4

KEY

a) to describe something that will happen in the future based on present knowledge or experience
b) to change from one form/state into a different one
c) a factory or other place where an industrial process is carried out
d) dependence on something

FOCUS ON FUNCTIONS EXERCISE 5

KEY

1 c
2 Suggested examples:
  a) decision: I'll take the job./I won't take the job.
  b) promise: I'll phone tomorrow./I won't be late.
  c) prediction: It'll be cold tonight./It won't snow today.
3 To make a prediction based on clear present evidence. (first example)
  To state an intention, something which is already decided/planned. (second example)

TAPESCRIPT

KEVIN: What will our sources of energy be fifty years from now?

JOHN: Fifty years from now we will see a completely different energy picture.

DISCOVERING LANGUAGE EXERCISE 6

KEY

1 a 3 b 2 c 1
2 & 3 certain: will definitely + infinitive without to
  probable: I think ... will + infinitive without to
  possible: may/might + infinitive without to

TAPESCRIPT

One.
I think we may come back to nuclear power. Perhaps it's dangerous at the moment but scientists might be able to design ways of making nuclear power stations safer in the future.
Two.
I think we'll have to change the way we live soon – unless they find far more sources of oil and gas.
Three.
You and I . . . all of us . . . we know that fossil fuels will definitely run out in twenty, thirty years. But we do nothing about it.

READING: A LITERARY EXTRACT EXERCISE 8
Background note
Ben Elton (1962 –) is one of the most famous alternative comedians in the UK. He writes satirical plays and novels, and has also acted in productions of his books.

背景知识注释:
本·埃尔顿 (1962-) 是目前英国两位最著名的喜剧作家之一。他专写讽刺性的剧本和小说,同时还在根据他的作品改编的剧目中扮演角色。

KEY

1 a) He thinks people are irresponsible and they do not think of the future.
  b) Politicians are only concerned with staying in power and finding short-term solutions.
2 The earth will die.
3 No, she doesn't. She calls his idea 'a pathetic generalisation'.
Note: *Man* is used in the first paragraph of the extract to refer to the human race/human beings.

SUGGESTED KEY

It's always been dying, ever since *the human race* began to take from it more than *it* needed. I tell you, Rosalie, Earth as we know it is finished, because *human beings* rule it and *they* are incapable of acting responsibly! Of thinking of anything other than the short-term.

Survival

Focus	SKILLS
TOPIC • Life in a closed ecosystem **GRAMMAR** • Articles	• Reading: an article • Listening: a radio news item • Writing: a paragraph **VOCABULARY DEVELOPMENT** • The prefix *self-* • Compound adjectives (time and size)

语法知识:

冠词 a 和 the 的区别:

1. the 具有"确定"的含义。在我们说 the car, the girls 等等 的时候,可能有两种情况:(1) 听话的人已知道我们所指 的是哪一辆汽车,或哪几位姑娘;(2) 我们正是要告诉听话的 人我们指的是哪一辆汽车,或哪几位姑娘。

   I had trouble with the car this morning.
   今天早上,车出了毛病。(指我自己的车。)

   Those are the girls who live next door.
   这就是住在隔壁的姑娘们。

   the 可以与 sun, moon, stars 等词连用。如果我们说 the sun, 指哪一个太阳是很清楚的(因为只有一个太阳)。如果我们 说 the stars, 意思也同样是清楚的(因为只有天上这些星 星)。

2. 如所指物为"不确定"的,我们一般用不定冠词a/an。我 们如果说 Pass me a piece of bread (请递给我一片面包), 这面包是不确定的,因为这可能是几块面包中的任何一 块。

   Shall we go and see a film?
   咱们去看场电影好吗?

3. 初次提到某件可数事物时,一般用不定冠词,因为听者对这 件事物一无所知。再次提到这一事物时,就用定冠词,因为 这时听话的人已经知道所指的事物了。

   A man came up to a policeman and asked him a question. The policeman didn't understand the question, so he asked the man to repeat it.
   有一个人走到警察面前,问了一个问题。警察没有听懂这 个问题,于是要他重说一遍。

4. 在若干常用词语中,介词后面可以省略冠词:

   to school　去学校　　　at school　在学校
   from school　从学校　　in/to class　在／去上课
   to/at/from university/college　去／在／从大学／学院
   to/in/into/from church 去／在／进入／出教堂
   to/in/into/out of prison/hospital/bed 去(上)／在／进入／离开监狱／医院／床
   to/at/from work 去上班／在上班／下班
   to/at sea 出海／在海上
   to/in/from town 进城／在城里／从城里回来
   at/from home 在家／离家
   for breakfast 早饭
   at lunch 吃午饭
   to dinner　(请人)吃饭
   at night 晚上
   by car/bus/bicycle/plane/train/boat 乘小汽车／公共汽车／自行车／飞机／火车／船
   on foot 步行
   to go to sleep 入睡
   如果上述词语与冠词连用,就有特殊的含义。
   He's in prison. 他在监狱里。(是一个犯人)
   He's in the prison. 他在监狱。(可能是去参观)

5. 在泛指广播和电视的时候,不用冠词。

   It's easier to write plays for television than for radio.
   写电视剧要比写广播剧容易一些。

在 listen to the radio 和 on the radio 这些词组里,要用冠 词,但在 watch television 和 on television (或 on TV) 这些 词组里,就不用冠词了。

6. 在提及到乐器的词组里,通常要用定冠词。如 play the guitar, learn the piano 等。

7. 疾病的名称一般是不可数的,用的时候不加冠词。

   I think I've got measles.
   我大概得了麻疹。
   我们可以说 a cold,例如 I've got a cold (我感冒了),但 是在 to catch (a) cold 这个词组里,有人不用冠词。
   我们可以说 a headache,但在英国英语中,其它疼痛 (toothache, earache 等)是不可数名词,不用冠词。

8. 在提及数字的词组中,要用不定冠词。例如:

   a hundred, a thousand, a million 等。
   It'll cost about a hundred pounds.
   那大概要花一百英镑。

9. 在表示季节时,我们可以说 spring (春天),也可以说 the spring,可以说 in summer (在夏天),也可以说 in the summer,等等。用不用冠词,意思没有什么区别。在 in the fall (美国英语,在秋天)这个词组里,一般要用冠词。

10. 在某些结构中,表示人们身份的名称不用冠词。

    They elected George chairman.
    他们选乔治当主席。

11. 船舶名称要用定冠词。

    The Queen Mary 玛丽女王号

12. 在用 country, sea, seaside 和 mountains 等词时,即使不专 指哪一个海,哪一座山,一般也要用 the。

    I love the mountains, but I hate the sea.
    我喜欢山,但是讨厌海。

13. 特指的江、河、海洋、山脉、群岛、地区、沙漠等通常用定 冠词 the。

    the Atlantic 大西洋　　　　the Himalayas 喜马拉雅山脉
    the West Indies 西印度群岛　the Rhine 莱茵河

14. 在洲、省、州、城镇、街道、湖泊、国家等地名前一般不用 冠词:

    Africa 非洲　　　　　　　Texas 得克萨斯
    Oxford 牛津　　　　　　　High Street 哈伊街
    Lake Windermere 温德米尔湖　Brazil 巴西

    国家名称包含普通名词者需用定冠词:
    The People's Republic of China 中华人民共和国
    The United States of America 美利坚合众国
    单独的山名一般不用冠词。例如:
    Snowdon 斯诺登山

READING EXERCISE 2
Background note

Star Trek is a popular American science fiction television programme made in the 1960s and 70s. It is about the adventures of a space ship (the *Starship Enterprise)* and its crew. There are also several *Star Trek* films. The title of the article is a pun on the commentary which introduced every episode of the TV series. The narrator always says 'to boldly go where no man has gone before'. Here the pun is on *grow*, instead of *go*, and *glasshouse enterprise* rather than the name of the ship *Starship Enterprise*.

背景知识注释:

《星球旅行》(Star Trek) 是 60 和 70 年代美国拍摄的一部十分受观众欢迎的科幻电视节目。它描述了太空探险号 (the Starship Enterprise) 宇宙飞船和它的宇航员的历险。同时也拍摄了几部《星球旅行》电影。本文的题目是一个双关语,因为有广播报导了这一电视连续剧的每一集节目。广播主持人总是说"勇敢地踏上了人类从未去过的地方。"这儿的双关语是用了"成长"(grow),而不是"去"(go),称太空探险号飞船为"玻璃房探险号"(glasshouse enterprise)。

KEY

Location: the Arizona desert

Sponsor: Edward Bass, a multibillionaire from Texas

Cost: $150 million

Number of inhabitants: 4 men, 4 women, 3,800 plant and animal species

Duties: to plant, harvest and process their food on a farm and to conduct experiments

Aims: to establish a self-sufficient community which might be used in a spaceship or on another planet; it would be a self-sufficient 'world' floating in space

Criticisms of the project: Many scientists don't believe it will work because the only closed ecosystem which has survived more than a few days was smaller than a football and only contained shrimp and plant life. Mr Bass has also been criticised for wanting to make money rather than do serious scientific research because he opened the biosphere to the public before the experiment began, and hundreds of tourists paid to visit the place every day, all buying souvenirs.

DEVELOPING VOCABULARY EXERCISE 4

KEY

1 a) Mr Bass himself (self-titled)
  b) the community itself (self-sustaining)
2 The subject does the 'action' itself, with no help from anybody else.
3 self-locking – the case locks itself
  self-taught – the expert taught him/herself
  self-defence – you can defend yourself
4 a) a self-cleaning oven   b) a self-employed woman
  c) a self-portrait

Note: The prefix *self-* can have two meanings, i.e. *a self-destruct mechanism* = a mechanism that destroys itself, but *a self-service restaurant* = a restaurant in which you serve yourself.

EXERCISE 5

KEY

1 a) two-year (study)   b) half-acre (farm)
2 The form of a compound adjective using a time period or size is:

ARTICLE	+ NUMBER	+ -	+ TIME PERIOD/SIZE	+ NOUN
a	two	-	year	study
a	half	-	acre	farm

The time period is always singular, e.g. not a two-years study. The main stress falls on the number.

3 a) a two-metre high wall   b) a ten-day sale
  c) an eight-storey house   d) a hundred-year lease

DISCOVERING LANGUAGE EXERCISE 6

KEY

1 a) four men and four women, the eight, the team, the crew members, Abigail Alling, a US marine biologist
  b) a giant hi-tech greenhouse, the $150 million structure, the so-called Biosphere 2, the seven-storey glass and metal structure
2 a) the people – no article, the biosphere – a/an
  b) the
  c) no article (but note the so-called Biosphere 2, where the article is needed because of the adjective)
3 The definite article is used when referring to information that is shared by the writer and reader. It is also used when there is only one of something, e.g. the Arizona desert.
  The indefinite article is used when information is introduced for the first time (a self-sustaining community) or when the noun belongs to a class of things (a US marine biologist).
  No article is used when talking about things generally rather than specifically. It applies to both countable (mainstream scientists) and uncountable nouns (algae).

  With proper names, the definite article may be used or omitted depending on whether it is part of the name or not, e.g. *Star Trek*, The Tower of London.

EXERCISE 7

KEY

1 the (specific – the only one)
2 a (one of a group not mentioned before)
3, 4, 5 no article (these items in general)
6 the (refers back to Edward Bass)
7 no article (general)
8 a (first time mentioned)
9 the (second time mentioned)
10 a (first time mentioned)
11 no article (in general)
12 the (refers back to cowboy boots)

🔲 LISTENING EXERCISES 8 AND 9

KEY

Successes: The people have lived in an enclosed system for 18 months longer than anyone else has done. They have managed to grow most of their food and to recycle their waste and water.

Failures: Oxygen had to be pumped into the system because the atmosphere was too thin and it was making the people ill. They did not manage to grow all their crops successfully, either.

TAPESCRIPT

BROADCASTER

Almost two years ago four men and four women sealed themselves into a giant greenhouse in the Arizona desert to see whether they could survive in a completely enclosed environment. Next month, the 'Biospherians', as they have become known, will return to a mixed reception in the real world. They believe the enterprise has been a success. They've lived in an enclosed system for eighteen months longer than the previous record, held by Russian astronauts; most of their food has been grown inside the complex and all their waste and water has been successfully recycled.

But many experts have dismissed the project as having no scientific value since oxygen was pumped into the biosphere at the beginning of the year. Crew members were suffering from altitude sickness at the time because the atmosphere was so thin.

For Sally Silverstone, who has been responsible for food systems inside the biosphere, the low oxygen levels and crop failures have been the most frustrating part of the whole experience. She has often wanted to give it all up when she sees crops that have taken so much work to grow ruined. But she would do it all again. When I spoke to her on the phone last night she told me she had no plans to take a holiday on her 24 months salary after 're-entry' but will be spending the next few months preparing a new crew to continue the work in the Biosphere.

SPEAKING: Discussions

Focus

TOPICS
- Genetic engineering
- Life in the future

FUNCTIONS
- Asking for explanations
- Introducing examples
- Interrupting
- Talking about certainties, probability, possibility, plans and ideas

SKILLS
- Listening: a conversation
- Speaking: discussion

SPEECH PATTERNS
- Using intonation to allow or prevent interruption

GETTING STARTED EXERCISE 1

KEY

a pessimist

🔲 LISTENING EXERCISE 2

KEY

1 They are discussing medical research and genetic engineering; specifically research that will eradicate disease and allow parents to choose the exact qualities of their unborn child.
2 Pat is a pessimist. Alan is an optimist. Tim hasn't really thought about the subject. Sara hasn't thought about the subject in great depth.

EXERCISE 3

KEY

1 Pat (science creates problems . . . line 8)
2 Sara (I intend to have the perfect child . . . lines 29 – 30)
3 Tim (What's wrong with that? line 21)
4 Alan (. . . perhaps that won't happen, lines 36 – 37).

EXERCISE 4

Note: Your students may be confused between *to affect* and *an effect*. *Affect* /əˈfekt/ is the usual verb which means to change or influence someone or something, e.g. *Smoking can affect your health*. *Effect* /ɪˈfekt/ is the noun, e.g. *Smoking can have a bad effect on your health*.

注释: 学生可能混淆 affect 和 effect 这两个词。Affect 是动词, 表示改变或影响某人或某事。例如: Smoking can affect your health. 抽烟影响健康。Effect 是名词。例如: Smoking can have a bad effect on your health. 抽烟对你的健康有坏影响。

KEY

1 they (line 5) – geneticists
  them (line 9) – problems
  what (line 14) – uses
  they (line 15) – geneticists
  ones (line 17) – genes
  that (line 21) – identifying genes that affect appearance, sex and intelligence
  that (line 33) – the possibility that the world will be full of perfect people with too many men
2 a) cut down (line 3)   b) to get rid of (line 5)
  c) hereditary (line 5)   d) finding out about (line 15)
  e) affects (line 20)   f) variety (line 35)

FOCUS ON FUNCTIONS EXERCISE 5

KEY

1 a) What do you mean?   Like what?   What's wrong with that?
   b) Well, take (+ noun/-ing) genetic engineering, for instance.
   c) Wait a moment.   Just let me tell you something.
2 a) Just a minute ... Hang on ... Let me say something ... Listen ...
   b) Excuse me ... Could I ask a question? Forgive me, but, ... I'm sorry to interrupt, but ...

SPEECH PATTERNS EXERCISE 6

KEY AND TAPESCRIPT

PAT: Well, they're not only finding out about genes that cause diseases ... they're also going to know which ones give us the colour of our hair and eyes. They already know how to create male or female children ... and now they're saying there might be a gene that affects intelligence!

2 When the intonation falls: this signals that the speaker has finished an 'idea'.
3 By rising before a pause to show that the speaker wants to continue speaking.

FOCUS ON FUNCTIONS EXERCISE 7

KEY

	QUESTION 1	QUESTION 2
A They might (may)	+ infinitive without *to*	= a possibility
B You can't be certain	*that* + clause	= a certainty
C Science is likely	infinitive with *to*	= a probability
D I intend	infinitive with *to*	= a definite plan
E I'm (not) thinking of	+ -*ing* form	= an idea that is/is not being considered

EXERCISE 8

KEY

1 having/taking   2 to do   3 to be   4 write/speak
5 be able   6 that they will (be able to) give you

Progress check: Units 10 – 11

Grammar and functions

EXERCISE 1

1 were able   2 couldn't   3 will be able   4 can
5 managed   6 can't   7 could   8 will be able

EXERCISE 2

1 a) We'll definitely move next year.
   b) We're certain/It is certain that we'll move next year.
2 a) The company is likely to leave London.
   b) It is probable that the company will leave London.
3 a) It may move to Manchester.
   b) It is possible that it will move to Manchester.

EXERCISE 3

1 C   2 B   3 D   4 E   5 A

EXERCISE 4

1 When are you going to have the party?
   What time will it start?
2 Where are you going to have it?
   Will it be big enough?
3 How many people are you going to invite?
   Will you invite your parents?
4 How much food are you going to provide?/How much food is there going to be?
   Will it be home-made?
5 Are you going to have a band?
   Will you have your stereo outside?

EXERCISE 5

Tony had his party and **the** party was a great success. There wasn't a band, but there was music and **people** danced to it. Tony's neighbours came too, so the party did not move into **the** house until after midnight. **Tony's** friends were still enjoying themselves at three in **the** morning.

Vocabulary

EXERCISE 6

1 irresponsible people
2 a self-governing organisation
3 an unpleasant experience
4 a ten-mile race
5 self-sacrificing parents
6 an incorrect answer
7 self-educated people
8 a ten-pound note
9 an impractical solution
10 a dissatisfied customer

12 *News and views*

Making news

Focus	SKILLS
TOPIC • Newspapers GRAMMAR • Past simple passive	• Listening: a monologue • Reading: newspaper extracts

语法知识:

动词的被动语态形式是由动词 to be 的不同时态构成的，后面跟过去分词。

I wasn't invited, but I've come anyway.

我没有被邀请，但我还是来了。

Has Mary been told?

通知玛丽了吗？

不是所有动词都有被动形式。不及物动词不能用于被动语态，因为它们后面不能跟宾语，所以没有东西可充当被动式动词的主语。某些及物动词至少在一些意义上不能用于被动语态。大多数这样的动词是"静态"动词（表示状态而不是动作，而且常常没有进行时形式）。例如：fit, have, lack, resemble, suit 等。

They had a nice house. 他们有一幢好房子。（但不能说: A nice house is had ... ）

GETTING STARTED EXERCISE 1

KEY

1 critic /ˈkrɪtɪk/
2 foreign correspondent /ˈfɒrən kɒrɪˌspɒndənt/
3 editor /ˈedɪtə/
4 journalist/reporter /ˈdʒɜːnəlɪst/, /rɪˈpɔːtə/
5 sub-editor /ˈsʌbedɪtə/
6 cartoonist /kɑːˈtuːnɪst/
7 deputy editor /ˌdepjətɪ ˈedɪtə/

Documentary

LISTENING EXERCISE 2

KEY

1 c
2 open answers

EXERCISES 3 AND 4

KEY FOR EXERCISE 3

1 eight and a magazine
2 news, news review, business, classified advertising, art section, books, colour supplement/magazine

KEY FOR EXERCISE 4

1 a broadsheet newspaper

2 It is the newspaper's comment on the most important event of the week and is written by the editor.
3 the business section
4 *The Sunday Times* is the only newspaper in Britain to have a separate book section.

TAPESCRIPT

IVAN FALLON

It's a, a broadsheet newspaper, as opposed to what we call a tabloid newspaper. This is twice the size of a tabloid newspaper. It's a very large newspaper. It's a multi-section newspaper: we've got eight sections and a magazine. The, the first section, which is obviously the most important section, is what we call the News section; and in that we have home news, we have foreign news and we have analysis of some of the major events of the week. After that, we have more specialised sections; we have what we call the News Review section. This section includes what we call the 'leader', which is the newspaper's editorial comment – our views on what is the most important event of the week. This one here, which will be personally written by the editor – which is an attack on the government's economic policy and its employment policies – this is our contribution to the political debate.

The third section, which is the section I edit, is Business. This is a section that is amazingly well-read; it's the classified advertising section – classified advertisements being specialist small advertisements. Quite often they're tiny ads.

Now we get to our tabloid sections. We have effectively three tabloid sections. The reason for this is that we found that not everybody likes the large broadsheet format. We've developed this particular section as our art section. The book section (we call section six) was the first time in Britain that this type of separate section had been done on books, and is still the only separate book section in Britain. We also have a colour magazine, colour supplement, which is very much a separate part of the paper.

READING EXERCISE 5

KEY

1 11   2 The Culture   3 classified advertising
4 open answers

EXERCISE 6

KEY

a – 2   b – 7   c – 3   d – 9   e – 4   f – 5   g – 10
h – 1   i – 6   j – 8

EXERCISE 7

KEY

1 Magazine (Funday Times)   2 Business/Personal Finance   3 Style and Travel   4 News   5 Books

6 Magazine   7 Culture   8 Sport   9 Appointments
10 News Review

DISCOVERING LANGUAGE EXERCISE 8

KEY

1 subject + past of verb *to be (was/were)* + past participle
2 Hopes of an advance . . . ***were raised*** yesterday . . . The first convoy of evacuees . . .***was cancelled*** yesterday . . .
3 Because we don't know the subject/the subject is not important.

EXERCISE 9

Note: The **Chancellor of the Exchequer** is the British government minister in charge of finance. The work of the Chancellor is the same as that of the Finance Minister in many other countries. In some countries, e.g. Germany, the **Chancellor** is the chief minister of the government, equivalent to the Prime Minister in Britain.

注释：财政大臣是英国政府主管金融财政的部长。财政大臣的职责与许多国家的财政部长一样。在一些国家里（例如德国），Chancellor 是政府首席部长（总理），相当于英国的首相。

KEY

1 The South of England was hit by a storm yesterday.
2 The factory gates were locked by the bosses yesterday.
3 A head teacher was sacked by the council yesterday.
4 New taxes were announced by the Chancellor yesterday.
5 Liverpool was beaten by Tottenham yesterday.
6 An 8-year-old boy was arrested yesterday.

EXERCISE 10

KEY

His clothes were torn and covered with mud. His bicycle was broken. His bag was broken. His books were taken from his bag and they were drenched/ruined. He was pushed/kicked/hit.

Reading habits

Focus	
TOPICS • Objectivity and bias in newspapers • Reading habits GRAMMAR • Present perfect passive	SKILLS • Reading: comments for a survey • Listening: short interviews • Speaking: discussion, an interview • Writing: a report SPEECH PATTERNS • Word stress

READING EXERCISE 2

KEY

TYPE OF NEWSPAPER	FREQUENCY	REASONS FOR CHOOSING THIS NEWSPAPER	REASONS FOR NOT CHOOSING OTHERS
serious	Sundays	articles written by well-informed journalists	hates sensational stories and chequebook journalism

EXERCISE 3

KEY

1 b   2 b   3 c   4 c

🔊 LISTENING EXERCISES 4 AND 5

KEY FOR EXERCISE 4

TYPE OF NEWSPAPER	FREQUENCY	REASONS FOR CHOOSING THIS NEWSPAPER	REASONS FOR NOT CHOOSING OTHERS
1 none			all biased, most are right-wing
2 tabloid	daily	likes human interest stories and quick to read	they are depressing
3 quality tabloid	Sundays 2–3 times a week	good summary of world events stories about royal family	too respectful

KEY FOR EXERCISE 5

1 not politically biased
2 stories about people rather than wars and disasters
3 the British royal family, e.g. the Queen, Princess Diana; examples of other 'public figures': politicians, film stars, pop musicians

TAPESCRIPT

One.
I don't read newspapers at all. They're all biased. They've been taken over by companies who have political interests. And in Britain most of the papers are right-wing. There are one or two that I think are a bit more objective, but I'd rather listen to the radio.
Two.
Actually, I don't want to know about the details of all the troubles in the world – it's all so depressing. I like

human interest stories – stories about people, not wars and disasters. So my daily newspaper is a tabloid. The other good thing about a tabloid is that you can read it in ten minutes on the way to work. That's all I want.
Three.

I get a so-called quality newspaper on Sundays because it gives me a good summary of world events, but I also buy tabloids two or three times a week. People are critical of the pictures and stories of the royals and their problems that have been published in the tabloids, but I don't agree. Why shouldn't we know how they're spending our money? They're public figures, and only the tabloids give you the details – the other papers are too respectful.

💬 SPEECH PATTERNS EXERCISE 6

KEY AND TAPESCRIPT

1 sen'sation   edu'cation   infor'mation   imagi'nation
  e'dition   pro'duction
2 The syllable before the *-tion* ending carries the main stress, e.g. con'dition, 'station, radi'ation, centrali'sation.

DISCOVERING LANGUAGE EXERCISE 7

KEY

1 the subject + auxiliary *has/have* + *been* + past participle
2 In A and C the passive is used because the subject of the active is long and complex. In B it is used because the subject is obvious or unimportant.
3 an indefinite time in the past, no specific time
4 It is used to talk about an action that happened in the past but we don't know when, or an action that began in the past and continues into the present. It is used when the subject is unknown, unimportant, obvious, or long and complex.

EXERCISE 8

KEY

1 were flooded   2 were overturned   3 were torn
4 were killed   5 was blown   6 were injured*
7 have been taken   8 have been damaged
9 have been given
*It is possible to use the present perfect here as the people are *still* injured. However, its position in the paragraph about *last* night makes the past simple more appropriate.

CREATIVE WRITING: Changing perspectives

Focus

TOPIC
• Perspectives on events

SKILLS
• Speaking: interpreting events
• Reading: a narrative

• Writing: a perspective on events; a newspaper report

VOCABULARY DEVELOPMENT
• Adjectives and their connotations

GETTING STARTED EXERCISE 1

Note: Brits is a colloquial expression for the British. **Aussies** is the equivalent for Australians.

注释：Brits 是口语用词，指英国人。Aussies 等于 Australians，指澳大利亚人。

KEY

1 Australia and Britain (Sydney/Brits)
2 the chance to host the Olympics in the year 2000
3 Australia
4 a) B   b) A
5 a) A   b) B
6 They probably felt delighted.
  Olympics delight as Aussie bid wins.
  This focuses on the fact that events can be reported differently depending on the perspective of the person who is doing the reporting.

SPEAKING EXERCISES 2 AND 3

KEY FOR EXERCISE 2

1 They are on a country road.
2 It's evening/night.
3 The elderly couple have broken down.
4 open answers

SUGGESTED KEY FOR EXERCISE 3

• The younger man is having trouble with his own car.
• The elderly couple have called for help on their mobile phone, and the younger man has been sent out by a local garage.
• The couple and the younger man have arranged a secret meeting here to discuss a 'shady' deal (something illegal).
• The couple are the younger man's parents. He has escaped from prison (in a stolen car). His parents have agreed to help him.

DEVELOPING VOCABULARY EXERCISE 4

KEY

YOUNGER MAN	THE ELDERLY COUPLE
huge, ugly-looking	did not seem to know what
looked terrifying	to do
long, greasy hair	lifted his arm weakly
filthy, torn jeans	unsteady steps backwards
big, black leather boots	looked nervously
holding a large metal tool	vulnerable
waved in the air	
walked slowly	

EXERCISE 5

SUGGESTED KEY

1 *a frightening person:* powerful /ˈpaʊəfəl/, muscular /ˈmʌskjələ/, angry /ˈæŋgri/, frightening /ˈfraɪtnɪŋ/, sinister /ˈsɪnɪstə/, evil /ˈiːvəl/, strong /strɒŋ/, loud /laʊd/, sharp /ʃɑːp/, aggressive /əˈgresɪv/, threatening /ˈθretnɪŋ/
a helpless person: powerless /ˈpaʊələs/, defenceless /dɪˈfensləs/, frightened /ˈfraɪtənd/, worried /ˈwʌrɪd/, harmless /ˈhɑːmləs/, innocent /ˈɪnəsənt/, feeble /ˈfiːbəl/, naive /naɪˈiːv/, weak /wiːk/, gentle /ˈdʒentl/, silent /ˈsaɪlənt/, terrified /ˈterɪfaɪd/, confused /kənˈfjuːzd/, trusting /ˈtrʌstɪŋ/, frail /freɪl/

2 a) *similar meanings:* powerful/muscular/strong, frightening/threatening, sinister/evil, loud/sharp/aggressive
   b) *opposite meanings:* powerless/weak/feeble/frail/ defenceless, frightened/worried/terrified, innocent/naive

3 a) muscular (arms), sinister (eyes), strong (legs), feeble/weak (grip), sharp (nose), frail (body)
   b) (feel) powerless/defenceless, frightened, worried, terrified, confused
   c) powerful, sinister, harmless, evil, innocent, strong, naive, weak, gentle, silent, trusting, aggressive
   d) angrily, worriedly, feebly, weakly, loudly, gently, aggressively
   e) angrily, weakly, silently, threateningly

4 a) a helpful person: kind, gentle, sympathetic, sensitive, understanding, practical
   b) a person who is not afraid: brave, courageous, heroic, assertive, confident, self-assured

Talkback

A critical eye

KEY

The picture shows: an early black-and-white television; a more modern colour television; closed-circuit cameras in shops, which monitor shoppers' actions; an early video game; a very modern flat-screen television with external speakers and an interactive text display on the screen, which allows the viewer to choose items from a 'menu'.

13 *On show*

But is it art?

Focus	
TOPICS	SKILLS
• Exhibitions	• Reading: an article
• Art	• Writing: describing an experience
GRAMMAR	
• Past perfect simple	VOCABULARY DEVELOPMENT
	• *Go* + adjective

语法知识：

过去完成时由 had + 过去分词构成。如果我们正在说的已经是过去的事，要说到比这再早一些时候发生的事情，就用过去完成时。只有存在"第二个"过去或者比所说的过去还要早的情况，才能用过去完成时。

I could see from his face that he had received bad news.
从他的脸上我可以看出他接到坏消息了。

过去完成时常常用在间接引语里，放在像 said, told, asked, explained, thought, wondered 等过去时动词的后面。这个时态指的是在谈话或产生想法之前已经发生了的事情。

I thought I'd sent the cheque a week before.
我以为我一周前就已把支票寄去了。

READING EXERCISE 2

KEY

1 Because her work of art was thrown away by mistake.
2 suggested vocabulary:
   a) shiny, red, wobbly
   b) cultural, stimulating, interesting

EXERCISE 4

KEY

1 It was 34 red jellies on 17 plates, laid out in the shape of an arc on the floor of the art gallery.
2 four days
3 The officer in charge of the art gallery removed it because he thought that it was left-over food from a party.
4 It represented the human body and what happens to it after death.

5 a) jewel-like, very fresh, shiny and red
   b) dull and mouldy, smelling badly

DEVELOPING VOCABULARY EXERCISE 5

KEY

1 it goes grey   2 it goes bad   3 you go bald
4 it goes red   5 you go blind

DISCOVERING LANGUAGE EXERCISE 6

KEY

1 subject + past of verb *to have (had)* + past participle
2 b) happened first. The past perfect is used to describe an action which happened before another one in the past.
3 before the jellies were thrown away

Extra practice

The students work in pairs. Read the story (below) out to them, pausing after each prompt for them to write down their answers. Encourage them to be imaginative. Check as a class and write the correct answers on the board, creating a skeleton story. The students then write down what they can remember of the rest of the story. At the end of the activity, the students should have the whole text in their notebooks.

A bad evening
Peter arranged to meet Simon at 7.30. He arrived at 8.00 but Simon ... (*had already left*). So he went to a restaurant alone and ordered a meal. Twenty minutes later the waiter told him that he couldn't have a meal because the chef ... (*had walked out*). He went to another restaurant and had a meal. At the end of the meal, he put his hand in his pocket to take out his wallet and found that he ... (*had left it at home*).

He washed dishes for a couple of hours and then walked home. When he arrived, the front door was open and everything was in a mess because there ... (*had been a burglary*). Even worse, he couldn't phone the police because the burglar ... (*had taken the phone*). He went to his neighbour's house and used his phone. It took the police an hour to come because their car ... (*had broken down*). Peter went to bed after the police ... (*had gone*) and he didn't get up the next day!

EXERCISE 7

KEY

1 Because someone had thrown away the jellies.
2 She had planned the sculpture.
3 She had displayed them on 17 plates in the shape of an arc on the floor of the museum.

4 The museum official had thrown them away.
5 Because they were mouldy and smelling badly, and there was no sign.

The cartoon is making the point that, if jellies can be considered art, then anything can, even someone's dinner.

Learning experiences

Focus	SKILLS
TOPIC • Museums FUNCTIONS • Asking for and giving directions	• Reading: an extract from a brochure • Listening: two short talks • Speaking: a short talk, a role play

GETTING STARTED EXERCISES 1 AND 2

KEY FOR EXERCISE 1

1 a) C   b) A   c) B
2 open answers

Documentary

LISTENING EXERCISE 3

KEY

1 The Science Museum, London (Picture A)
2 It's the first of its type in the country.
3 60
4 a) school parties of pupils and teachers
   b) families
5 It's fun and visitors learn a lot from visiting it.

TAPESCRIPT

PRESENTER
Listen to Alison Porter, a curator at the Science Museum in London, talking about a particular gallery in the museum. Answer the questions in your Students' Book.
ALISON PORTER
Launch Pad is, is our largest interactive gallery and, when it opened, it was the first of its type in this country. It has a number of exhibits – as many as 60 different interactive exhibits are, are in the gallery.

During the week you'll see school parties using the gallery with their teachers, but at weekends families come and the whole lot join in. I find the museum a very stimulating place: it's a place that's both a lot of fun and it's also a big learning environment. And I think that anyone that visits here will go away feeling that they've learned something from the experience, and they can take that away with them.

READING EXERCISES 4 AND 5

KEY FOR EXERCISE 5

1 The International Red Cross/Red Crescent. It helps people who are victims of military conflicts or natural disasters. It was started in 1863.
2 There are two symbols because the cross, although not a religious symbol, was potentially offensive to Muslims.

EXERCISE 6

KEY

1 1859
2 There is a red cross on the Swiss flag. This was originally used as a tribute to the Swiss founder of the organisation.
3 Because it is also a symbol of Christianity.
4 open answers

🔲 LISTENING EXERCISE 7

KEY

1 a d e g h
2 original agreements setting up the organisation, computer records of current prisoners of war, sculptures of groups of people who are faceless and have their hands tied

TAPESCRIPT

Well, there are a lot of fascinating exhibits. There's the Wall of Time. This gives details of major world events – wars, conflicts and natural disasters – from 1863 to the present day, and gives details of the Red Cross involvement in each one. It's also got a display of the original agreements setting up the organisation and examples of medical kits carried by volunteers.

Then you can see the kind of records the organisation used to keep. One room has a card index system of all prisoners of war held in thirty-eight different countries during the First World War. Of course nowadays records are held on computers . . . and visitors can consult these if they wish.

There are also films of Red Cross and Red Crescent volunteers in action in the First and Second World Wars, and of its activities in peacetime fighting disease and helping victims of earthquakes, volcanic eruptions, and other natural events. The work of the organisation today is shown on a row of television screens.

Another interesting feature is that a number of sculptures of groups of people are placed throughout the museum. The people are faceless and their hands are tied. They represent the misery of casualties of war and oppressive peace throughout the world.

It's a surprising museum. It leaves you feeling depressed at the amazing scale of human suffering but it's also encouraging to think that organisations like the Red Cross and Red Crescent exist and are dedicated to peace.

🔲 EXERCISE 8

KEY

exhibit /ekˈsɪbɪt/– display
fascinating – interesting
card index system – records
wars – conflicts
work – activities
casualties /ˈkæʒəltɪz/– victims
misery – suffering

FOCUS ON FUNCTIONS EXERCISE 10

KEY

a) café, restaurant   b) cloakroom
c) souvenir shop, gift shop   d) toilet   e) exit
f) ticket office   g) restaurant   h) picnic area
i) souvenir shop, gift shop   j) library

SPEAKING: Responding to pictures

Focus		SKILLS
TOPICS		• Reading: a poem
• Children		• Speaking: a discussion, a description
• Childhood		• Listening: a description
FUNCTIONS		
• Describing location		**VOCABULARY DEVELOPMENT**
• Describing people		• Words with similar and opposite meanings
• Interpreting/inferring from pictures		
• Expressing opinions and feelings		
• Hesitating		

READING: A POEM EXERCISE 1

KEY

Neither of them can look after themselves independently and both need help from other people.

EXERCISE 2

KEY

a) beads   b) veins   c) banisters   d) grown-up
e) grand   f) wrinkles

EXERCISE 3

KEY

1 She thought they tried to be 'grand'.
2 In a negative way, both physically and in terms of character.

3 She saw a friend of her great-aunt Etty (who was very old) try to pick up the beads from a necklace. It had broken and the beads were all over the floor. She had difficulty picking them up.
4 Open answers, e.g. no teeth, lose their hair, can't concentrate on things, can't walk far, short memory, physically weak.

EXERCISE 5

KEY

1 It is likely to be something on children in different countries.
2 open answers

📼 FOCUS ON FUNCTIONS EXERCISE 6

KEY

1 C
2 a) describing location
  b) describing people (what you can see)
  c) describing people (what you can guess)
  d) expressing opinions
  e) expressing your feelings

📼 EXERCISE 7

KEY

let me see   you know   now   so   (also 'sounds', e.g. er, um)

TAPESCRIPT

There are two children in the picture, one in the middle at the front and another behind. I don't know if they're boys or girls . . . They look very similar . . . perhaps they're twins. They've both got blond hair and . . . er . . . let me see . . . I think they've got blue eyes. They're wearing nappies and they're covered in paint! They've got paint everywhere – in their hair, round their faces and all over their bodies. I guess they're . . . em . . . probably at a nursery . . . you know . . . or a playschool, or somewhere like that.

They look about one or two years old . . . and they're obviously well-fed . . . they're both quite plump. Now . . . I think they're probably European . . . or American perhaps. Actually, in my opinion this isn't a very natural photograph. It's not likely that both kids would have the dishes on their heads at the same time . . . so . . . I think the photographer arranged the photo.

Anyway, it's a nice picture. The kids look happy and innocent and secure . . . and, you know, they don't seem to have a care in the world. It makes me feel good to see kids so happy and relaxed.

DEVELOPING VOCABULARY EXERCISE 8

SUGGESTED KEY

rich – wealthy (similar)
hungry – well-fed (opposite)
helpless – vulnerable /ˈvʌlnərəbl/ (similar)
neglected – spoilt (opposite)
healthy – ill (opposite)
cheerful – hopeful (similar)
carefree – relaxed (similar)
miserable – cheerful (opposite)
arrogant – proud (similar)

Progress check: Units 12–13

Grammar and functions

EXERCISE 1

1 What did Jane and Tom do on the last day?
  They shopped at the Friendship store.
  Had they shopped there before?
  No, they hadn't.
2 What did Lenny and Pam do on the last day?
  They walked on the Great Wall.
  Had they walked on it before?
  Yes, they had.
3 What did Tricia do on the last day?
  She visited the Summer Palace.
  Had she visited it before?
  No, she hadn't.

EXERCISE 2

1 when   2 just   3 never   4 already   5 yet
6 by the time

EXERCISE 3

1 Manchester United was beaten by Benfica last night.
2 The pitch was invaded (by the crowd) after the match.
3 Shop windows were smashed outside the ground.
4 Over a hundred people have been arrested.
5 More than fifty people have been treated for minor injuries.

EXERCISE 4

1 of, take   2 on, first   3 there, use
4 How, to the right of, downstairs

Vocabulary

1 Yes, but it's gone cold.
2 He has gone deaf.
3 Everything has gone wrong today.
4 She went crazy.
5 Yes, the nail has gone black.

14 *In touch*

Door to door

Focus	SKILLS
TOPIC • Couriers GRAMMAR • Reported speech	• Reading: an article, an advertisement • Listening: a conversation • Speaking: reporting an experience

语法知识:

引用某人的话或想法的一种方法是用"间接引语"结构。在这种情况下,我们不是一字不改地引用原话来说明某人的意思,而是把他的意思跟我们自己的句子更紧密地联系起来(比如用 that 或 whether 引导)。

She explained that she hadn't recognized me.
她解释说她没有认出我来。

如果"表示引语的动词"是过去时(如 she said; I thought; we wondered; John wanted to know),我们一般不用原说话者所使用的时态,而要改为"过去"(因为我们不是跟说话者在同一个时间里说话)。试比较:

直接引语	间接引语
一般现在时	一般过去时
现在进行时	过去进行时
一般过去时	过去完成时
现在完成时	过去完成时
过去进行时	过去进行时或过去完成进行时

有时当我们转述别人所讲的事情时,这些事情仍然未变,则所用动词也不变。

"I'm only 18." "我才 18 岁。"

She told me the other day that she's only 18.
几天前她对我说她才 18 岁。

READING EXERCISE 2

KEY

1 by going as a courier, accompanying documents or packages

EXERCISE 3

KEY

express delivery (defined above)

fare-paying passenger (a passenger who pays the correct fare)
customs procedures (checks at the airport for customs purposes)
return flight (a flight back from a destination)
full-fare ticket (a ticket with no reduction in price)
hand luggage (luggage which can be carried onto an aircraft)
last-minute bargain (a flight/ticket which is very cheap because it is bought only a day or two in advance)

EXERCISE 4

KEY

1 A courier takes urgent documents or parcels from one place to another. In this article, by plane from Britain to abroad.
2 It's a cheap way of travelling abroad and you can get some good bargains. The work doesn't take long and it's easy to do.
3 If you want to use the flight to go on holiday, there isn't a wide choice of destinations and the dates of the flights are fixed so you can't choose when you want to go. If you're travelling as a couple, you usually have to go on separate days because only one courier is needed a day. If you want to travel together, you have to pay for one full airfare. You also have to check in at least two hours in advance and be dressed quite smartly. The representative may only arrive at the airport thirty minutes before the flight, which can make you nervous, and you may have to carry a lot of hand luggage.

EXERCISE 5

KEY

POSITIVE	NEGATIVE
excellent (meal) steady (nerves) /ˈstedi/ fast (service) low (cost) reliable /rəˈlaɪəbl/ (staff, courier)	restricted (range of destinations) /rɪˈstrɪktɪd/ fixed (dates) /ˈfɪkst/ anti-social (courier flight) /ˈænti ˌsəʊʃl/ smart (clothes) /smɑːt/ expensive (staff)

LISTENING EXERCISE 6
Background note

In Britain, some companies operate a **freephone** system. This means that the company will pay for the telephone calls made to it, especially in answer to an advertisement. In the UK the code for these free numbers is 0800.

背景知识注释:

在英国,一些公司实施免费拨打电话制度。这意味着公司将支付拨打给它的电话费用,特别是对公司广告所作的应答电话。在英国,免费拨打电话号码的代码为 0800。

KEY

1 ICS
2 It delivers parcels to any destination at any time of the day.
3 Freephone 0800 121213
4 They will come and collect your parcel and organise for delivery to New York.

🔲 EXERCISES 7 AND 8

KEY FOR EXERCISE 7

1 books
2 from near Oxford to New York
3 less than 24 hours (if the parcel catches the midday flight)

KEY FOR EXERCISE 8

1 the invoice address; the pick-up address if it's different and a contact phone number; the name, address, phone number of the person the parcel is being sent to; the weight, dimensions and value of the parcel and a full description of the contents
2 check the contents and take the invoices
3 the sender's address and the destination address; the contents of the parcel; the weight and value of the contents; a sentence saying that the information is true and the sender's signature

TAPESCRIPT

EMPLOYEE: ICS. Good morning. How can I help you?

WOMAN: Good morning. I'm thinking of sending a parcel to New York by courier next week. Can you tell me what the procedure is, please?

EMPLOYEE: Certainly. When you ring us, we need the following information: the invoice address – that's probably your own address, isn't it? – and then the pick-up address if that's different. And a contact phone number...

WOMAN: Just a moment... I'm taking notes. Phone number...right.

EMPLOYEE: Then we need the full name, address and phone number of the person you're sending the parcel to.

WOMAN: OK. Anything else?

EMPLOYEE: Yes, the weight and dimensions of the parcel – that's height, width and length...and the value of the goods...and a full description.

WOMAN: ...value...description...

EMPLOYEE: Yes, but don't seal the parcel. You need to leave it open so that the driver can check the contents when he collects it. After the recent bombing, the airlines said that we'd have to check all parcels; they told us we had to do it.

WOMAN: Fine. Now, how long will the parcel take to get to New York?

EMPLOYEE: One to two working days. Just a moment, I'll check with the manager. Yes, the manager says that there are daily flights at midday. If your parcel catches that flight, it'll arrive within 24 hours.

WOMAN: Good. I live near Oxford. What time would you need to collect from here in order to catch the midday flight?

EMPLOYEE: Near Oxford...What's your postcode?

WOMAN: OX7.

EMPLOYEE: Just a moment. Yes, the manager says we can collect the parcel from you at 10.15 on the morning of the flight. Now, there's one more thing. What are you planning to send?

WOMAN: Books.

EMPLOYEE: In that case, you'll need to write an invoice for customs. We need six copies, and the driver will take them from you when he collects the parcel. On the invoice you need to write your address and the destination. Then you must say what's in the parcel, and specify the weight and the value. Write a sentence declaring that the information is true, and sign each copy.

WOMAN: ...sign each copy. Right. Thank you very much. You've been very helpful.

EMPLOYEE: Not at all. Goodbye.

WOMAN: Goodbye.

DISCOVERING LANGUAGE EXERCISE 9

KEY

1 no
2 Actual words: You will have to check all parcels; you have to do it.
  you – we, would – will, have to – had to
3 The reporting verb in A is in the present and so the tenses do not have to change. In B the reporting verbs (said, told) are in the past and so the tenses change.

4

DIRECT SPEECH	REPORTED SPEECH
present simple	past simple
present progressive (are waiting)	past progressive (were waiting)
past simple (left)	past simple/past perfect (had left)
present perfect (have phoned)	past perfect (had phoned)
past perfect	past perfect
will, can, may	would, could, might
must/have to	had to
would, could, should, might	would, could, should, might

Emergency

Focus	
TOPIC • Emergency services **FUNCTIONS** • Telephoning	**SKILLS** • Listening: an interview, phone calls, a conversation • Speaking: a role play, a report of a conversation

GETTING STARTED EXERCISE 1
Background note
A **paramedic** is a member of an ambulance crew who is trained to give medical attention in emergencies, but is not a doctor or a nurse. **Ambulance drivers** or **ambulance men and women** are trained to assist people, but they do not have the same level of medical training as paramedics.

背景知识注释:
医务辅助人员是救护工作人员之一。他们接受过在紧急情况下进行医疗护理的培训。但他们不是医生，也不是护士。救护车驾驶员或救护人员受过救护病人的培训。但是他们并未接受过与医务辅助人员相等水平的医疗培训。

Documentary

LISTENING EXERCISE 2

KEY
1 London Ambulance control room
2 999
3 a British Telecom operator
4 what service you need
5 the exact address of where the ambulance needs to go
6 why the caller needs an ambulance
7 advice
8 that a patient had walked to the surgery with a heart attack
9 to get an ambulance to the address as quickly as possible

TAPESCRIPT
PRESENTER: Listen to an interview with Michelle Redfern, who answers the telephone at the London Ambulance control room.
Part One.
MICHELLE: When somebody dials 999, they'll speak to a British Telecom operator. The operator will ask the caller what service they require – either police, fire brigade or ambulance. The caller then will say 'ambulance service'.
Hello, London Ambulance. Can I help you?... 12 Edrick House, Page Street. Hold on, what, what area of London is that?

The information you, you want from a caller is the correct location – that is very important – you need to know where they actually are.
OPERATOR 2: *Yes, what's the address, sir? 36 where? Larch – L-A-R-C-H Close. And that's where sir? SW... South West 12?*
MICHELLE: Once you've actually got the correct information, you then find what is wrong, why they need an ambulance, if you can help – give advice over the phone, and get an ambulance to them as soon as you can.
Part Two.
PRESENTER: The interviewer asked about calls that Michelle had received that day. Listen.
MICHELLE: I received a call from a doctor's receptionist and she said to me that a patient had walked to the surgery with a heart attack.
London Ambulance. Can I help you?... Is this an immediate ambulance?... And the doctor's name?
The main aim of my job is to help the public and get them the ambulance there as soon as we can – in the shortest time possible.

SPEAKING EXERCISE 3
Background note
Fulham is an area in West London.

KEY AND TAPESCRIPT
BT: Emergency Services. Fire, police or ambulance?
C: Ambulance, please.
BT: I'm putting you through.
AS: Hello. London Ambulance. Can I help you?
C: Yes, there's been an accident ... a boy has been hit by a car.
AS: Can you tell me exactly where you are?
C: In Fulham, outside number... 44, Birchfield Avenue.
AS: Can you spell that, please?
C: B-I-R-C-H-F-I-E-L-D. Birchfield.
AS: Is the boy conscious?
C: Yes, but he's losing a lot of blood ...
AS: OK, we'll have an ambulance there as soon as possible.

COMPARING CULTURES EXERCISE 5
Background note
In Britain, it is very unusual to answer a private phone by saying your name. People usually say the telephone number, although it is becoming increasingly common just to say *Hello*. It is likely that a business call would be answered with either the company or the individual's name.

背景知识注释:

在英国,通常在接私人电话时,人们不报出自己的姓名。人们通常报出自己的电话号码。现在越来越普遍的是人们只说一声"喂"。然而如果是公事电话,那么接电话人一般报出公司或自己的名字。

KEY FOR EXERCISE 2

The man is Mr Naylor. *Speaking* is the short form of *This is (Mr Naylor) speaking.*

📼 FOCUS ON FUNCTIONS EXERCISE 6

KEY

a 2, 7   b 5   c 4   d 1   e 3   f 6

EXERCISE 7

SUGGESTED KEY

1 Hang on. Wait a moment, please.
2 I'm afraid she's out at the moment/she's at lunch/she's away today/she's on holiday/she's very busy at the moment.
3 Because he's making an offer (*I'll put you through*) and a promise (*I'll let her know*). They are both spontaneous decisions.
4 Shall I give her a message? Shall I tell her you rang? Shall I take a message? Could you call back later? Why don't you try again in half an hour?

📼 LISTENING EXERCISE 9

KEY

1 Ms Bayliss's son
2 He had a motorbike accident.
3 He's broken his leg and has got cuts and bruises on other parts of his body.
4 No.
5 24 hours
6 She's going to the hospital and is taking some pyjamas and toiletries for him.

TAPESCRIPT

MS BAYLISS: Hello, this is Helen Bayliss.
NURSE: Hello, Ms Bayliss. I'm phoning from the hospital.
MS BAYLISS: Yes? Has something happened?
NURSE: I'm afraid your son has had an accident.
MS BAYLISS: What? Oh no . . . is he hurt?
NURSE: Well, he's broken his leg and he's got cuts and bruises to other parts of his body.
MS BAYLISS: But he's OK?
NURSE: Yes, he's seen a doctor and he's going to be all right.
MS BAYLISS: Oh thank goodness! Can I come in and see him?

NURSE: Yes, of course. And perhaps you could bring some pyjamas and toiletries in for him.
MS BAYLISS: So you're going to keep him in overnight?
NURSE: Yes, but don't worry. It's just routine. The doctor wants to keep him under observation for 24 hours. He wants to make sure there are no problems.
MS BAYLISS: What do you mean? What kind of problems?
NURSE: Oh . . . nothing. Really, it's just routine after an accident like this.
MS BAYLISS: I see. But . . . what happened? What sort of accident was it?
NURSE: He was riding his motorbike and was hit by a car.
MS BAYLISS: Oh . . . that motorbike . . . I've told him to be careful! When did this happen?
NURSE: Oh . . . about three hours ago.
MS BAYLISS: Right . . . er, well, I'll go and get some things for him and I'll be there as soon as I can.
NURSE: That'll be fine. Come to reception in Accident and Emergency.
MS BAYLISS: Right. I'll be there soon. Thanks. Goodbye.
NURSE: Goodbye.

WRITING: Short reports

Focus

TOPIC
• Home computers

FUNCTIONS
• Use of tenses in sections of a report

SKILLS
• Reading: an article, a report
• Speaking: a survey
• Writing: a report of a survey

VOCABULARY DEVELOPMENT
• Verbs for reporting results

READING EXERCISES 2 AND 3

KEY

The headline means that girls want to use computers to learn but boys like playing with computers (*Game Boys*).The reasons for the differences are that fewer girls have computers at home and girls are less enthusiastic about Information Technology at school. They are more interested in the word-processing and database functions of computers than in video games.

The possible effects are that because girls are less confident and enthusiastic about IT at school, they do not do as well as boys do in this subject. Consequently, there are fewer girls than boys on higher-level computing courses in colleges and schools.

DEVELOPING VOCABULARY EXERCISE 4

KEY

1 Shows

2

SUGGESTS	SHOWS
indicate	prove
imply	demonstrate
provide some evidence	confirm
	provide conclusive evidence

READING EXERCISE 5

KEY

1 1 introduction   2 carrying out the survey
  3 results 1   4 results 2   5 conclusion

2 How old are you?
  Do you have computers at home?
  How much time do you spend on your computer in
  an average week?
  What do you use your computer for?

3 The bars show the number of young people who use
  computers for:
  a playing games   b word processing
  c keeping addresses and telephone numbers/keeping
  a diary   d programming   e studying
  f consulting databases   g anything else
  The mistake is that both f and g should read zero, not
  one as in the case of f.

FOCUS ON FUNCTIONS EXERCISE 6

KEY

a) present simple (*this report presents, the results
   indicate, the chart gives*)
b) past simple (e.g. *we questioned, asked*)
c) past simple and past progressive (*used, they were
   learning*)
d) present simple (e.g. *show, dislike*)
reported speech

EXERCISE 7

KEY

young people   teenagers   users   people questioned
respondents   14–18 year olds   adolescents
The writer uses different expressions to give variety.

15 *A change of scene*

Preparing to leave

> **Focus**
>
> TOPICS
> • Preparations for travel
> • Working holidays
>
> GRAMMAR
> • First conditional
> • Conjunctions: *if, unless,
>   when, as soon as*
>
> SKILLS
> • Speaking: discussion
> • Reading: an article

语法知识:

在 if 所引导的条件状语从句中,通常不用 will 或 shall 而用现
在时表示将来。

If I have enough money next year, I will go to Japan.
如果明年我钱够的话,我将去日本。

conjunctions: if, unless, when, as soon as

与 if 所引导的条件状语相同,unless, when, as soon as 所引导
的状语从句表示将来也用现在时。

Unless I finish my homework, I will not go to bed.
除非我做完功课,否则我不去睡觉。

When I finish my homework, I will go to bed.
我做完功课后,就上床睡觉。

As soon as I finish my homework, I will go to bed.
我一做完功课就上床睡觉。

GETTING STARTED EXERCISE 1

SUGGESTED KEY

Documents: passport, visas, travellers' cheques, credit
  cards, driving licence, letter of employment, 'register'
  of all the teenagers, receipts for camera and other
  equipment
Clothes: waterproof clothes, socks, hat, sunglasses,
  T-shirts, underwear, jeans, jumpers, swimsuit, shorts,
  pyjamas (or nightwear in general)
Toiletries and medicines: toothpaste, deodorant,
  antiseptic cream, hairbrush, aspirin, plasters, first-aid
  kit, suntan cream, soap, malaria tablets, nail scissors,
  shampoo
Equipment: whistle, rucksack, torch, Swiss army knife,
  water bottle, maps, town plans
Books: maps, phrase books, novels, puzzle books
Other things: pack of cards, pens and pencils, paper,
  sweets, alarm clock, camera and film, walkman, spare
  travel bag

READING EXERCISES 2 AND 3

SUGGESTED KEY

1 open answers
2 tour guide, waiter, au pair, grape picker, shop
  assistant, hotel staff, interpreter, courier

SPEAKING EXERCISE 4

SUGGESTED KEY

1 Advantages: work abroad, practise foreign languages, visit interesting places, earn money, no special qualifications needed, good experience
Disadvantages: work long hours with little free time, low pay, can be difficult work (e.g. the courier in Europe), no job security
2 patient, resourceful, calm, fond of children, caring, energetic, keen on sports
3 mostly seasonal jobs: grape picking in France, working on a campsite, working on American summer camps for children, being a nanny, washing dishes in restaurants, being a holiday courier for American children visiting Europe, working in a hotel

EXERCISE 5

KEY

1 work that is only available during a certain time of the year, e.g. jobs in shops at Christmas and during sales, summer jobs
2 the excitement of travel
3 short-term work which is not guaranteed and is needed for a special purpose, e.g. when a company wants to send out a lot of catalogues, it may employ casual labour to put the catalogues in the envelopes and stick on address labels; casual work does not depend on the time of year
4 the rights a worker has, as set down in the law
5 to lose your job because you have done something wrong, e.g. if you have been rude to a customer

DISCOVERING LANGUAGE EXERCISE 6

KEY

1 If you work as a nanny . . . , you'll have to speak Italian. (*lines 16–18*); You'll work if it's convenient for the company . . . (*lines 43–44*); If you don't work hard, or if your employer doesn't like you, you'll get the sack. (*lines 47–48*)
2 a) Unless you speak the language of the country well, there will be very few openings. (*lines 13–15*)
b) When you arrive to wash dishes in a restaurant in Paris, the owner will expect you to speak French. (*lines 18–20*); You'll have a job when the hotel . . . is busy. (*lines 41–43*)
c) As soon as the holiday season finishes, they'll get rid of you. (*lines 46–47*)
3 a) Both *if* and *unless* are used in conditional forms but *unless* means *if . . . not*, e.g. If you *don't speak* the language . . .
b) Both *when* and *as soon as* are time conjunctions. *As soon as* has the meaning of immediately. They follow the same form as conditional sentences, i.e.

when/ as soon as + present simple, *will* + infinitive without *to*.
4 There is no difference. If the sentence begins with *if*, *as soon as*, *when*, *unless*, there is usually a comma between the two clauses. If the conjunction is in the middle of the sentence, there is no comma.

EXERCISE 7

KEY AND TAPESCRIPT

MOTHER: You'll contact your employers immediately, won't you?
SON: Don't worry! I'll contact them as soon as I arrive.
MOTHER: Look after your bags in public places, won't you!
SON: Of course I'll look after my things if I'm in a public place!
MOTHER: Now, remember not to eat unwashed fruit.
SON: It's all right. I won't eat any fruit unless it's been washed.
MOTHER: And make sure you boil water in rural areas.
SON: Yes, yes. I'll boil the water if I'm outside the cities.
MOTHER: You won't forget to take your malaria tablets every morning?
SON: No. I'll take my malaria tablets when I get up.
MOTHER: And you'll let me know your new address . . . .
SON: Don't worry. I'll write to you when I find somewhere to live.
MOTHER: It says here that it's important to start learning the local language immediately.
SON: Yes, I'll start learning it as soon as I get there.
MOTHER: And you won't accept invitations from strangers, will you?
SON: No, Mother. I won't accept an invitation unless I know the person well. Now, is that it? Can I finish packing?

Getting away

Focus

TOPIC
• Holidays

GRAMMAR
• Indirect questions
• Embedded questions

FUNCTIONS
• Asking politely

SKILLS
• Speaking: discussion, role play
• Listening: a monologue, a conversation at a travel agent's
• Writing: form completion

语法知识：
间接疑问句不用直接疑问句常用的词序（助动词置于主语前），不用 do，也不用问号。
The doctor asked how I felt.

医生问我感觉如何 。

在间接疑问句中，问题前面如果没有疑问词（如 who, where, why），就用 if 或 whether 。

The bus driver asked if/whether I wanted to go to the town centre.

公共汽车司机问我是否要到市中心去 。

I don't know if/whether I can help you.

不知道我能不能帮你的忙 。

🔊 Documentary

📼 EXERCISE 4
Background note
Kuoni is a travel company in the UK which is well-known for its exotic holidays in quite distant places.

背景知识注释:

Kuoni 是英国的一家旅游公司 。它以安排去遥远的地方，让游客享受异国情调的假日而闻名 。

SUGGESTED KEY FOR EXERCISE 3

The photographs show: a hotel in Kurumba in the Maldives, a Caribbean cruise, the Nile and a Thai temple. The students should infer that the company specialises in long-haul, luxury holidays.

KEY FOR EXERCISE 4

1 c   2 b   3 a
4 Thailand, Hong Kong, Barbados, Egypt

TAPESCRIPT

MARIA PAUL

Kuoni offer some very exciting tours. We go to some quite unusual places that possibly people haven't heard too much about: little islands like Koh Samui. We go to places like the Maldive Islands – lots of different islands with very different characters, different things to offer everyone, everything from a particularly deluxe island like Kurumba down to tiny little islands with just very basic accommodation, restaurants with sand floors, that type of thing. And then there's completely the other end of the market: Princess cruises around the Caribbean, the ultimate in, in luxury. The customers can ask for absolutely anything and we'll provide it, but I think probably the most popular destinations overall are Thailand, Hong Kong, Barbados in the Caribbean, although other Caribbean islands also do very well; Egypt is always popular – a lot of people are fascinated by Egypt and the culture; the Nile cruises in particular are always very full. Working in the travel business, you find that every day is different. You don't know what the client is going to want. When they come through the door, they could ask for anything. They could ask

for a rail ticket or they could ask for the most wonderful exotic holiday.

📼 EXERCISE 5

KEY

1 Egypt   2 two weeks   3 by plane and boat
4 three (Cairo, Aswan, Luxor)
5 in the morning (10.45)
6 a, b and c

📼 WRITING EXERCISE 6

KEY

BOOKING FORM

Names of travellers

Surname	Initials	Mr/Ms
Porter	S. A.	Ms
Porter	G. L.	Mr

Telephone
Home: 386775   Office: 625903

Travel details

Outward	Date	16th April	Time	10.45
	From	Heathrow	To	Cairo
	Airline	Egyptair		
Return	Date	30th April	Time	20.50
	From	Cairo	To	Heathrow
	Airline	Egyptair		

Accommodation

Hotel name   Pullman Maadi Towers
           Number of nights 4
Boat name   Osiris
           Number of nights 10

Deposit   2   people at £100 each = £ 200
Insurance   2   people at 41 each = £ 82

TAPESCRIPT

AGENT: OK, so I'll just go through that with you again. Two adults. Two weeks in Egypt from April 16th to the 30th. Flying from Heathrow to Cairo and back with Egyptair. Two nights at the Pullman Maadi Towers in Cairo, then by air to Aswan. A ten-day Nile cruise from Aswan, through Luxor and back to Cairo, followed by two more nights at the Pullman Maadi Towers, flying back to Heathrow on the 30th. And you're in a double room at the hotel and a double cabin on the boat, which is the Osiris.

MAN: Yes, that's right. Um ... do you know what the flight times are?

AGENT: The outward flight from London is, er ... let me see. Yes , 10.45 ... in the morning. And the return is ... I think it's early evening ... yes ... 20.50. Ten to nine in the evening. Local time, that is.

MAN: Right, that's fine.

WOMAN: Oh ... sorry. I can't remember what you include in the price. Is it all meals or just breakfast?

AGENT: Yes, it's full board ... so all meals ... and transfer from the airport to your hotel ... everything's included.

WOMAN: Good.

AGENT: Now, can you tell me if you need travel insurance?

WOMAN: Yes, yes we do.

AGENT: OK, well, that's an extra £41 each. Is that OK?

MAN: Well, there's no choice is there? I mean, we have to have it, don't we?

AGENT: Yes, I'm afraid so.

MAN: Well, all right then.

The conversation stops here in Exercise 5, but continues in Exercise 6.

AGENT: OK. I wonder if I could I take some details now. Right. Could I have your names, please?

WOMAN: The surname's Porter. P-O-R-T-E-R.

AGENT: And your initials, please?

WOMAN: S.A.

AGENT: S.A? And yours, sir?

MAN: G.L.

AGENT: G.L. Right. And can I have your telephone number, please?

MAN: Our home number's 386775.

AGENT: 386775. And do you have a work number – in case I need to contact you during the day?

WOMAN: You can put mine down. It's 625903.

AGENT: Fine. OK. If you could just sign the booking form here. Thank you. And I need a deposit.

WOMAN: Now? Er ... can I pay by credit card?

AGENT: Yes, of course, then we'll call you when the tickets arrive from the airline. I don't know when you want to pay the balance of the holiday price, but we will need it before we can issue the tickets. Perhaps you could come into the office some time next week?

WOMAN: Er ... yes, I mean, that should be OK.

FOCUS ON FUNCTIONS EXERCISE 7

KEY

1 to be more polite or formal
2 A: What are the flight times?
The verb and subject are inverted.
B: Do you need travel insurance?

If is replaced by the auxiliary verb *do*.

3 C, D and E are statements grammatically and questions functionally.

4 C: What do you include in the price?
The auxiliary *do* is added before the subject.
D: Could I take some details now?
The verb and subject are inverted.
E: When do you want to pay the balance of the holiday price?
The auxiliary *do* is added before the subject.

EXERCISE 8

KEY

1 Could you tell me how much this costs?
I wonder how much this costs.
2 Can you tell me which airport you want to leave from?
I don't know which airport you want to leave from.
3 Can you tell me if you want a luxury hotel?
I can't remember if you want a luxury hotel.
4 Do you know if you would like to hire a car?
I don't know if you would like to hire a car.
5 Do you know when you will be able to pay the balance?
I wonder when you will be able to pay the balance.

SPEAKING EXERCISE 9

SUGGESTED KEY

1 Can you tell me what time the outward flight is?
Of course, your plane leaves at 16.45.
2 And I can't remember how much the baggage allowance is.
You're allowed to take 20 kg.
3 Can you tell me when I arrive in Bangkok?
Yes, certainly. The flight arrives at 05.50.
4 And I don't know the name of the hotel.
It's the Hotel Caledonian.
5 I can't remember how many days the holiday is.
It's 10 nights.
6 I wonder if I've already booked the car.
No, you haven't.
7 Do you know if I have all the meals at the hotel?
No, you only have breakfast at the hotel.
8 Can you tell me if I need a visa?
Yes, you do.

CREATIVE WRITING: Persuasion

> **Focus**
>
> TOPICS
> - Persuasion
> - Travel brochures
>
> SKILLS
> - Reading: persuasive and neutral texts
> - Writing: a persuasive description
>
> VOCABULARY DEVELOPMENT
> - Adjectives: degrees of intensity

READING EXERCISE 1

KEY

1 A is from an advertisement.
  B is from information on the back of a guidebook.
  C is from a personal letter.
2 A is written for potential travellers.
  B is written for travellers to a particular country.
  C is written for a friend.
3 A is to sell flights/to persuade people to fly with a particular airline.
  B is to persuade people to buy the guidebook.
  C is to invite the friend to come to the island on holiday.

EXERCISE 3

KEY

1 someone who was thinking of going on holiday to Bali
2 The qualities mentioned are: the weather, landscape and scenery, resorts and facilities, buildings, entertainment, local handicrafts and culture, the local people.

EXERCISE 4

KEY

1 The order is b, d, a, c.
2 Nature/climate: radiant sunshine, fresh/verdant landscape, beautiful sunsets, dazzling white beaches, stunning mountain scenery, clear blue lake and sea waters
  Modern facilities: vibrant coastal resorts, high-quality modern hotels, exciting nightlife, bright lights
  Local culture: traditional way of life, spectacular temples, colourful ritual dances, shadow puppet plays, vast range of handicrafts, the strength and vitality of the local culture
  People: friendly and gentle Balinese will welcome you

EXERCISE 5

SUGGESTED KEY

1 b) green c) bright d) pretty e) busy
  f) interesting g) big
2 a) has b) see c) has existed d) enjoy
3 a) countryside b) pleasures c) pleasant place
4 is blessed, temples, ritual, paradise (This question focuses on the general 'feeling' given to the text.)
5 Visitors from all over the world come to see the mountains, lakes and sea. In the busy coastal resorts you will find modern hotels and nightlife, but you will also find a traditional way of life which has existed for a long time. You can see the temples, ritual dances, shadow puppet plays and a big range of handicrafts, which show you something of the local culture. The Balinese people will welcome you and you will want to return to this place.

Note: The final sentence above could be considered opinion only and could be omitted.

DEVELOPING VOCABULARY EXERCISE 6

SUGGESTED KEY

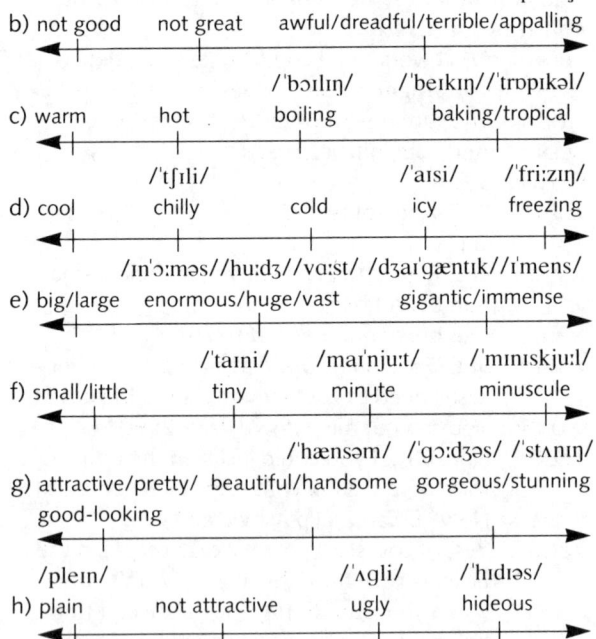

b) not good   not great   /'ɔːfl/ /'dredfəl/   /ə'pɔːlɪŋ/
                            awful/dreadful/terrible/appalling

c) warm   hot   /'bɔɪlɪŋ/   /'beɪkɪŋ//'trɒpɪkəl/
                  boiling   baking/tropical

d) cool   /'tʃɪli/   cold   /'aɪsi/   /'friːzɪŋ/
           chilly   icy   freezing

e) big/large   enormous/huge/vast   gigantic/immense
                /ɪn'ɔːməs//hjuːdʒ//vɑːst/ /dʒaɪ'gæntɪk//ɪ'mens/

f) small/little   /'taɪni/   /maɪ'njuːt/   /'mɪnɪskjuːl/
                   tiny   minute   minuscule

g) attractive/pretty/   /'hænsəm/ /'gɔːdʒəs/ /'stʌnɪŋ/
   good-looking   beautiful/handsome   gorgeous/stunning

h) plain   /pleɪn/   not attractive   /'ʌgli/   /'hɪdɪəs/
                                         ugly   hideous

Progress check:
Units 14–15

●●●●●●●●●●●●●●●●●●●●●●●●●●●●●●●●●●●●

Grammar and functions

EXERCISE 1
1 When   2 arrive   3 If   4 isn't   5 will show
6 unless   7 If   8 take   9 unless   10 want   11 When
12 will need   13 as soon as   14 can   15 if
16 will get

EXERCISE 2
RECEPTIONIST: Good afternoon, can I help you?
CLIENT: Good afternoon. Can I speak to the
          manager, please?/Could you put me
          through to the manager, please?
RECEPTIONIST: I'm afraid he's not here at the moment.
CLIENT: Could you tell me where he is, please?
CLIENT: Could you put me through to his
          secretary, please?
RECEPTIONIST: Could you tell me your name, please?
RECEPTIONIST: Could you just hold the line, please?

EXERCISE 3
1 I don't know if it rains in Morocco in January.
2 I can't remember if you need a visa.
3 I wonder which the best hotel in Casablanca is.
4 I'll find out where you can ski.
5 I know which currency they use.

EXERCISE 4
1 She said it rained in coastal areas but that it didn't
  rain much in Marrakesh.
2 She said that she had asked the embassy about visas.
3 Last week she asked me if I could check the hotels in
  my guidebook the next day.
4 She told me that they were planning to go into the
  Atlas Mountains.
5 Before they left, she said that she would ask about
  money because they probably couldn't buy
  Moroccan dirhams here.

Vocabulary

EXERCISE 5
1 How big was the desert?
2 How interesting/nice/pretty were the mountains?
3 How bright was the sun?
4 How good was the food?
5 How cold was the sea?
6 How bad was your sunburn?

Teacher development tasks

Introduction

It is easy to be so concerned with the progress of our students that we forget about the importance of our own professional development. We owe it to our students, and especially to ourselves, to be open to new ideas and perspectives, to extend our own awareness of language, to learn from others and to re-evaluate our own procedures. For this reason, a number of worksheets for teacher development have been included here. We feel that they are potentially useful to a wide range of teachers, even those with considerable experience.

The content of these worksheets focuses on both language awareness and classroom practice. All are set in the context of challenges faced when working with intermediate learners.

The worksheet tasks are intended to raise awareness and provoke discussion. Notes on each of the worksheets are provided at the end of the Worksheet section, but these are our comments presented from our perspective and are presented as suggested answers only. You may find that the insights you gain from using the worksheets are different from and/or go beyond what is offered in the notes.

Who are the worksheets for?

1 INDIVIDUAL TEACHERS WORKING ALONE

Although it is always helpful to discuss ideas with other teachers, it is possible to work through the tasks on your own. You may then wish to look at the notes at the end of the section or turn to teachers' handbooks and resource books which focus specifically on areas that you would like to explore further.

2 TEACHER TRAINEES WORKING WITH A TEACHER TRAINER

The worksheets can be photocopied and can therefore be used freely as input or follow-up self-study materials for seminars or courses. Ways of working with the materials are described more fully under 3 below, but one possible procedure is to do the tasks in pairs or small groups and follow this up with a more general discussion led by the trainer.

3 GROUPS OF TEACHERS WHO MEET TO SHARE THEIR EXPERIENCE AND IDEAS

Teachers working in the same institution or geographical area often meet regularly to discuss theoretical and practical teaching issues. Meetings such as these are particularly useful to teachers who are relatively new to the profession and who can be helped by the greater experience of their colleagues. However, more experienced teachers will also find it helpful to reflect on their own knowledge and repertoire of techniques and may be surprised and stimulated by the quite different ideas of some of their colleagues.

You can work through a complete worksheet in a self-development session, or select tasks that are particularly relevant to the group. There are a number of possible ways of preparing for sessions. You can – individually – read and think about either the worksheets themselves or the general focus beforehand, and perhaps do some background reading so that you have more to contribute to the discussion. You can also think about and note down questions that you have in relation to the topics and any problems that you wish to raise at the meeting. Following the session, you may wish to read books and articles that your colleagues have recommended on the subject and these may well lead you to wider discussions on related topics.

Motivating intermediate learners

1 Think about the statement below. Do you agree with it? Why (not)? What arguments might be used by a teacher whose opinion is different from yours?

❝ Intermediate learners are more difficult to teach than beginners. ❞

2 Look at these comments by intermediate learners and their teachers.

1 Do you recognise the problems on the left as ones that intermediate learners often have? Do your students have these problems? If not, why not?
2 Do you share the frustrations of the teachers on the right? If not, why not?

A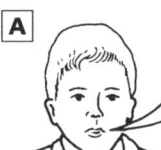
I've got no idea how much progress I'm making. There's always so much more to learn.

D
My classes are mixed ability. Some students are bored and others get left behind.

B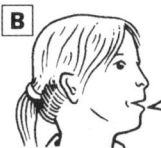
I've learnt a lot of different structures and functional phrases, but I don't really know when to use them.

E
Students enjoy communicative classes, but they have to take traditional exams which don't test these skills.

C
I feel silly when I speak because I make mistakes and I can't think of the right word.

F
At this level my students need to be exposed to a lot more English outside the class, but I haven't got time to encourage and monitor this.

3 Consider these suggestions for dealing with Problem A above. Are there others you can add? How could you try to overcome the other problems? Write the aim and two or three strategies for each one.

A *Aim:* to show students what they have achieved
Strategies:
Make learning aims clear.
Give frequent informal and formal tests with quick feedback.
Praise successful use of language rather than concentrating on correction.
Repeat activities done at the beginning of the year/course to show how much less challenging they have become.
When authentic texts have been understood, emphasise that they were written for native speakers.
Play tapes of real conversations to show that native speakers hesitate, struggle for words and misunderstand each other.

4 Use your experience to complete the following tasks.

1 List other motivational problems that learners may have:
   a) as a group   Example: *Each student in the class has different requirements.*
   b) individually   Example: *Particular needs are not satisfied.*
   Which of these are particularly true of intermediate students?
2 Note down additional problems that you have when teaching intermediate students.
   Example: *They use their first language even when they don't need to.*

5 Consider strategies for dealing with the problems that you identified in Exercise 4.

What's in a written text?

1 Read the texts below and discuss these questions.

1 Can you predict how each text is likely to continue in terms of content and overall organisation, grammar and vocabulary?
2 Can you predict more about some texts than others? Why?
3 Where might you find each text?

2 Consider the ways in which texts are organised.

1 Match these organising principles with the texts as you expect them to develop.
   a) problem – solution   d) background – instructions
   b) chronological sequence   e) setting – events
   c) question – answer
2 How would you expect these text types to be organised?
   a) a letter of complaint   b) a book review   c) an academic research report

A

SO WHAT WOULD £100,000 GET YOU TODAY?
Ten years ago, £100,000 would have bought you a three-bedroom flat in Kensington, an old rectory in the country or an island off Scotland. Four years later, when the market peaked in the last quarter of 1988, £100,000 was just below the average price for any house in Greater London and the South-East. That average has now fallen to £77,000. So what would £100,000 buy you today?

B

Canasta, a game of the Rummy family, originated in South America. In 1949 it spread like wildfire across America, and it has since become popular all over the world. The game is easy to learn and extremely exciting, but at the same time its strategy is sufficiently complex to interest serious card players.

C

If you are faced with a completely new garden, attached to a house that has just been built and consisting of a bare patch of soil studded with builders' rubble, your first instinct might be to rush off to the nearest garden centre and buy armfuls of plants to cover up the bareness. There are some things in life, however, that can't be rushed, and planning a garden is definitely one of them.

D

P. D. James was born in Oxford in 1920 and educated at Cambridge High School. From 1949 to 1968 she worked in the National Health Service as an administrator, and the experience she gained from her job helped her with the background for *Shroud for a Nightingale*, *The Black Tower*, and *A Mind to Murder*.

E

She might have been waiting for her lover. For three-quarters of an hour she had sat on the same high stool, half turned from the counter, watching the swing door. Behind her the ham sandwiches were piled under a glass dome, the urns gently steamed. As the door swung open, the smoke of engines silted in, grit on the skin and like copper on the tongue.

*See page 91 for sources of these texts

3 What is the overall purpose of Text A? What structural device does the writer use to achieve that purpose? What features of the text make it clear that this is a complete paragraph?

4 How do writers link clauses/sentences to make texts flow? Find examples of:

a) a noun phrase that refers back to another noun. (Text B)
b) a pronoun that refers back to a noun. (Text B)
c) an adverb that refers back to a particular time. (Text B)
d) a noun phrase that refers back to a much longer phrase. (Text A)

5 How formal is Text C? How does the writer achieve this level of (in)formality?

6 Compare Texts D and E.

1 In which text does the writer have a wider choice of possible lexical items? Why is this so?
2 What is the effect in Text E, the opening paragraph of a novel, of using these words rather than the possibilities given in brackets?
   a) lover (boyfriend)   b) same high stool (omit *same*)   c) swing door (omit *swing*)
   d) steamed (hissed)   e) silted (drifted)
3 Text D begins with 'P. D. James' and Text E begins with 'She'. What different effects do these choices create?

Look Ahead Intermediate (Hopkins/Potter) © Longman Group Ltd 1995

What's in a spoken text?

1 Without looking at the text on the right, list the features that you feel characterise spoken language.

2 The text is the transcription of a real conversation.

1 What is the relationship between the two speakers?
2 What is the purpose of their conversation?

3 Study the text in detail and answer these questions.

1 Which words frequently occur in subject position?
2 In what ways is the organisation different from that of a written text? Look, for example, at lines 10–16.
3 What features do you notice in lines:
  a) 8–10?
  b) 17, and 20–21?
  c) 39–41?
4 What is happening in lines 33–38?
5 What are the different meanings of the words *nice* and *dogmatic* as they are used between line 23 and the end? Are the meanings of these words constant throughout?
6 What do you notice about the range and choice of vocabulary?
7 To what extent are the speakers:
  a) transmitting information?
  b) being co-operative with each other?
  c) being clear and concise?
8 To what extent do you feel the features you have looked at in 1–7 above are characteristic of spoken texts in general?

4 Which of the features that you noted in Exercise 3 do you think are typical of a service encounter in a shop?

5 Look back at your list from Exercise 1. Do you want to amend it in any way?

NATHAN: We've fought for all this time for the simple reason that you refuse to allow me to have the space in my life to invite people over to stay here.

GLORIA: That's not true. I said I know I'm not -5 gonna feel like this forever – let me think about how I can deal with this. The way I have decided to deal with it is not to be here. I don't want to be here. I just don't want to be here. I -10 don't want to be around someone I'm totally uncomfortable with and I can't be myself around because everything he's gonna say … and I'm gonna be sitting there like – oh no, I can't -15 believe he said that!

NATHAN: Then why can't you … Gloria, I feel like your role in life is to react to things that happen around you and that's it, that's all you ever do. Why don't you, -20 I mean, come on, you know, why don't you talk to some people? For one thing you'll be less dogmatic … it'll be nice.

GLORIA: I don't want to be less dogmatic. I like -25 being dogmatic. I like thinking what I think. I like having strong opinions.

NATHAN: You don't have to change any of those things …

GLORIA: You want me to be nice and mellow -30 and polite and docile and I'm not willing to be your nice little wife …

NATHAN: Gloria. Nice, mellow, polite and docile are four words. Three of them mean similar things and one of them is not a -35 synonym. Docile … you don't have to be nice *and* docile … that's not the same thing.

GLORIA: Nathan, I'm not nice …

NATHAN: … polite and docile are different … -40

GLORIA: … I'm just not nice … and you're gonna have to just live with that if you're interested. I'm just not nice. That's not what I am. I don't walk around buttering people up because -45 I'm supposed to be this nice female type saying – 'Oh yes, let me get you some tea' … 'Would you like something to drink?' … I am not nice.

A If it doesn't rain, the crops will die.	D The crops died if it didn't rain.
B The crops wouldn't die if it rained.	E The crops won't die if it's rained.
C If it doesn't rain, the crops die.	F If it were to rain, the crops might not die.

1 Look at the sentences in the box above.

1 Add words and phrases to provide a context for each sentence.
2 Which structures are intermediate students likely to recognise?
3 Which ones can you name?
4 What do we teach intermediate students about the forms and uses of the structures you chose in question 2?
5 What mistakes do students make with the standard form of these structures?
6 Choose a structure from the list above that intermediate students are not usually taught. If a student asked about it, how would you explain its use?

2 Now study the sentences in detail.

1 Which sentences refer to situations which are:
   a) hypothetical?   b) past?   c) future?   d) general truths?
2 Which words or phrases can replace *if* or *if ... not* in these sentences?
3 In which sentences does *if* mean *whenever*?
4 What are the similarities and differences between:
   a) the basic zero conditional pattern in Sentence C and the variation in Sentence D?
   b) the basic first conditional pattern in Sentence A and the variation in Sentence E?
   c) the basic second conditional pattern in Sentence B and the variation in Sentence F?
5 Which other forms of the verb *rain* can replace:
   a) *doesn't rain* in Sentence A?   b) *rained* in Sentence B?
   What do these verb forms have in common? How does the meaning of the sentence change in each case?
6 Which auxiliaries can replace:
   a) *will* in Sentence A?   b) *wouldn't* in Sentence B?
   What is the effect of these different auxiliaries on the meaning of each sentence?
7 What generalisations can you make that can help you explain variations in the first and second conditional patterns that are taught to intermediate students?

3 Which of the functions in the box can describe the uses of Sentences 1–8 below? Can you generate other *if* sentences for each function? Practise saying them with appropriate intonation.

a) prediction   b) suggestion   c) threat   d) criticism   e) offer
f) explanation   g) negotiation   h) polite request

1 'If you do that once more ... !'
2 'If you could reduce the price, I'm sure we could come to some arrangement.'
3 'Go and lie down if you're tired.'
4 'If I could just ask you to wait for a few minutes ...'
5 'You can have it if you like.'
6 'If they were your own tools, you'd keep them clean.'
7 'Italy will win if Baggio is playing.'
8 'If you press that switch, the stage lights will come on.'

What's in a word?

1 What, in your view, does it mean to 'know' a word in your own or a foreign language? Note down the points that occur to you.

2 Consider the word *honest* as an example of a particular lexical item.

1 What part(s) of speech can it be?
2 Can we add any prefixes to it? Which one(s)?
3 Can we add any suffixes to it? Which one(s)?
4 Are there any features of spelling and pronunciation that you would need to bring to the attention of your students? What are they?
5 Is there a direct translation in your students' language(s)?

Answer the remaining questions with reference to the English word or, if you find this difficult, refer to the word that is most similar in meaning to *honest* in your language.

6 Are there other words that have similar meanings? Are the meanings and contexts of use exactly the same? If not, what are the differences?
7 Are there words that have an opposite meaning?
8 Are there any words that commonly occur with this word? (i.e. what sort of people and things can be *honest?*)
9 Can you think of any fixed phrases or idioms in which the word occurs?

3 Think about the way in which cultural connotations of words can differ.

1 Look at this statement: ❝ I was tried by an honest judge. ❞
   a) What exactly does *honest* mean in this case?
   b) What do we learn about the speaker's view of judges?
   c) Which of these characteristics do you associate with judges?
      serious   unpredictable   sympathetic   corruptible   wealthy   intelligent
      insecure   fair   pompous   privileged
      Do you know of other cultures in which the associations are different?

2 What are the connotations of these words and phrases to you, and to people generally in your culture?
   a) a cup of tea, e.g. *Will it normally have milk in it?*
   b) pride, e.g. *Is this a negative quality?*
   c) politicians, e.g. *Are they representative?*
   d) football fans, e.g. *Are they anti-social?*
   e) to shout, e.g. *Do you do this when you are happy?*

4 'Lexical items' (e.g. *honest*) are often distinguished from 'function words' (e.g. *because, for, can*).

1 Consider the word *so*.
   a) How many different structures can you think of in which it occurs?
   b) How many different meanings of *so* can you identify?
   c) When *so* is used as a conjunction, what are its main functions? Which other words have similar functions?

2 Some lexical items, like *reason*, can also have clear functions.
   a) Which words frequently occur in these positions?
      the reason . . . . this/that reason
   b) Contextualise each phrase in one or two sentences. In each case, does the phrase come before or after the reason that is given?

5 Do you want to add anything to the list you made in Exercise 1?

1 Look at these contexts of language use. For each context, list the language that you would expect to occur naturally in terms of:

a) the grammatical structures   b) the functional language

1 A group of people are having a discussion about life in the future. Some are optimistic and others are pessimistic.
2 A family are in a travel agent's to book a holiday. They are not sure of the kind of holiday they want.
3 Two people are standing on a mountain. Below them they can see a town, a number of villages and three or four beaches. Using their map, they are trying to work out what each place is called.

2 One function often suggests another which is likely to follow it in a conversation, e.g. greet – greet; give an opinion – agree/disagree. Which functions would you expect to follow these?

a) compliment –   e) complain –
b) invite –   f) express fear –
c) state a problem –   g) give permission –
d) ask for an explanation – h) offer –

3 Match the language areas on the left with contexts in which they are likely to occur and which would therefore be useful for language practice. Think of a sample sentence for each context that includes an example of the language area.

1 *used to* + infinitive
2 past progressive/past simple contrast
3 present perfect passive
4 present perfect progressive
5 reported speech
6 giving advice
7 stating obligations and prohibitions
8 talking about preferences
9 discussing plans and intentions
10 asking polite (indirect) questions

a) a conversation at a tourist office
b) an end-of-term party
c) a test on common road signs
d) a verbal report of a meeting with a famous person
e) an account of childhood
f) agreeing on a travel itinerary
g) a progress report on the building of a new house
h) an annual interview focusing on an employee's performance
i) a conversation with a friend who wants to overcome a phobia
j) an eye-witness's report of a crime

4 Can you think of a different context for each of the language areas in Exercise 3 – a context that would interest your particular students?

5 Consider different ways of establishing a context for language practice for your students.

1 Add to this list:
 • cards for role play (e.g. Exercise 3 a)
 • a simple instruction (e.g. Exercise 3 b: Find out your classmates' plans for the summer)
 • picture cards (e.g. Exercise 3 c)

2 Choose a way of introducing each of the contexts that you thought of in Exercise 4.

Notes on the Worksheets

Worksheet 1

Aim: to explore ways of dealing with problems of motivation

1 Examples:
(*Agree*) My knowledge and intuitions are constantly challenged in intermediate classes, so I have less confidence.
(*Disagree*) I don't need to prepare as many different activities as I do for beginners, because students have much more to contribute.

2 open answers

3 B *Aim:* to help students use language appropriately
Example strategies:
Within clear contexts, explore the effects of choosing different language patterns.
Record one group doing a freer practice activity; play it to the class, praise the strengths, and elicit phrasing that would make it even more communicatively effective.

C *Aim:* to give students the confidence to speak freely
Example strategies:
Be generous with praise, particularly with less confident students.
Teach students ways of rephrasing or paraphrasing if they can't find the exact words.

D *Aim:* to deal effectively with mixed-ability classes
Example strategies:
Seat more and less able students together, to work in pairs.
Set the better students more complex tasks based on the same input.

E *Aim:* to reconcile communicative methodology with traditional tests
Example strategies:
Dedicate one class in five to exam practice and techniques, so that students can confidently transfer their skills to other task types.
Ask students to work together on exam-type activities, so that the tasks become more communicative.

F *Aim:* to encourage students to learn independently
Example strategies:
Ask students to report to you and the class

(e.g. via notes on a wallchart) what they have read, listened to, written or talked about in English outside class in the last week.
Place a box in the classroom to which students add articles, advertisements, videos, etc. that they found interesting, so that others can borrow them.

4 1 a) Example: They can see no clear reason for learning English.
b) Example: Some students resent producing long pieces of writing in their own language – and even more so in English.
(A particular problem at intermediate level and above, where writing tasks are more substantial.)
2 Example: Giving feedback after group work is difficult because students have so much language to experiment with that they may have a wide range of problems and questions.

5 open answers

Worksheet 2

Aim: to highlight some key characteristics of written English

The texts on page **86** are from:
A *The Independent*, 23/7/94 (Anne Spackman)
B *The Pan Book of Card Games*, 1960 (Hubert Phillips)
C *Absentee Gardener Spain*, Anaya 1989 (Susan Pendleton)
D *The Children of Men*, Penguin 1994 (P. D. James)
E *England Made Me*, Heinemann 1935 (Graham Greene)

1 1 Open answers, but the following is a suggestion for Text A.
Content: current house prices in different parts of the country
Organisation: answers to the question at the end of paragraph 1 – one paragraph per area – and then a summary or recommendation about when and where to buy for the best price
Grammar: present tenses, first and second conditional structures, prepositional phrases (e.g. with country views, in a quiet area)
Vocabulary: types of property (e.g. semi-detached house), descriptive/evaluative adjectives (e.g. vast, pleasant), etc.
2 Possible answer:
It is easier to predict the development of a text if:

a) you are familiar with that text type.

b) the conventions that structure the text type are relatively inflexible (e.g. Text D).

c) there are explicit pointers to what is to follow (e.g. Text A).

3 A in a newspaper or magazine

B in a book on card games

C in a book/magazine for amateur gardeners

D on the dust jacket or before the title page of a novel

E in the body of a novel or short story

2 1 a) C   b) D   c) A   d) B   e) E

2 Possible answers:

a) general statement of complaint – detail of problem – suggested solution

b) information – introduction – critique – recommendation/summary of views

c) abstract (summary) – introduction – methods – results – discussion/conclusion – bibliography/notes – appendices

3 To introduce the main topic by providing background information. The structure of the paragraph makes it clear that it is complete. Having considered the recent history of price movements, the writer concludes this introduction by repeating the question (also the title of the article), which brings the reader forward to the present day. The use of the word *So* to introduce the question emphasises that the digression to the past has finished.

4 a) *a game* and *the game* refer back to *Canasta*

b) *it* and *its* also refer back to *Canasta*

c) *since* refers back to *1949*

d) *That average* refers back to *the average price for any house in Greater London and the South-East*

5 It is relatively informal – even chatty – for a book extract. Note, for example, the use of *you/your*, *are faced with*, *that has just been built*, *rush off*, *armfuls of plants*; the abbreviation *can't*; the structure of the paragraph, in which the writer takes into account the readers' first instincts before gently suggesting a more sensible alternative.

6 1 Text E, because of the creative potential of fiction

2 The effect of a writer's words obviously vary to a certain extent according to the experiences of the reader. For us the effects are as follows:

a) *lover* suggests secrecy, furtiveness and exoticism; *boyfriend* removes this layer of implied meaning and gives a feeling of openness and banality;

b) the use of *same* emphasises the fact that she has not moved during the time she has been in the cafe;

c) to add *swing* to the word *door* evokes possible sounds, e.g. creaking, and the possibility of a sudden entrance, rather like in a cowboy saloon; it also helps to build up a visual picture of the place;

d) *steamed* suggests something warm, homely and secure; *hissed* is a more violent word that evokes different, harsher sounds;

e) *silted* is an unexpected word here which gives the impression that the smoke-filled air has a certain solidity about it; this solidity is confirmed as the sentence continues. To use *came* would be to imply nothing beyond the fact that a cloud of smoke entered the room.

3 The subject of the text is specified immediately in Text D. Readers of her novel know that relevant biographical information will follow. In Text E, we are onlookers of a scene; the author creates interest in the woman by asking us to imagine who and what she is. As observers we do not know her name.

Worksheet 3

Aim: to highlight some key characteristics of spoken English

1 open answers

2 1 They are husband and wife.

2 To win a battle! Gloria does not want Nathan's friends staying in the house. Nathan wants her to welcome his friends.

3 1 *I, you* and *we*

2 The speech is not organised in paragraphs, nor are there necessarily identifiable, well-structured sentences. Punctuation has been imposed in the transcription, as far as possible, on a sequence of utterances. (Nathan's first speech is better formed, as if it has been planned in advance.)

3 a) Repetition of ideas and particular phrases.

b) Rephrasing ideas, breaking down and restarting utterances.

c) Gloria and Nathan are not listening to each other; both are determined to finish what they want to say, so they interrupt one another.

4 The focus is on language as they analyse and

negotiate the meaning of words.

5 *Nice* is positive in line 24 (it means *pleasant*) and is negative in lines 30 and 32 (*undemanding/submissive*). After that, Nathan uses it to mean *kind* in a positive sense, while Gloria continues to use it as a quality which stereotypes and degrades her as a woman. To Nathan, *dogmatic* means *assertive* in an intolerant sense; to Gloria, being assertive about your views is a positive quality.

6 The vocabulary is quite restricted in range and informal except when the focus is on language itself.

7 a) very little
  b) Nathan is fairly co-operative and open to discussion; Gloria is not.
  c) They are struggling to make themselves clear to a hostile listener; neither is concise.

8 Most features are typical of informal discussions, in which speakers are thinking about their ideas as they speak. However, most conversations are more co-operative than an argument.

4 The features are not at all typical unless the shopkeeper is well-known to you. Speech that is characteristic of service encounters tends to be clear, concise, formulaic and co-operative.

5 open answers

Worksheet 4

Aim: to demonstrate the creative potential of varying a basic structure

1 1 Examples:
  A ... doesn't rain *soon* ...
  B ... wouldn't die *so regularly* ...
  C ... doesn't rain *by May* ... die *each summer.*
  D ... died *each summer* ... didn't rain *by May.*
  E ... die *this year* ... rained *by the end of next week*
  F ... were to rain *soon* ... might not die *this year.*
  2–4 A (1st conditional) See *Look Ahead Intermediate* Students' Book p54.
  B (2nd conditional) See *Look Ahead Intermediate* Students' Book p54.
  C (zero conditional) If + present simple verb, present simple verb; used to state a general rule.
  5–6 open answers

2 1 a) B, F   b) D   c) A, (B), E, F   d) C, D

2 A *Unless* it rains ...
  B ... *as/so long as* it rained.
  C *When* it doesn't rain ... ; *Unless* it rains ...
  D ... *when* it didn't rain; ... *unless* it rained.
  E ... *as/so long as* it's rained; *provided/providing (that)* it's rained.
  F *Provided/providing that* it were to rain ...
3 C, D
4 Structures and meanings are similar, but:
  a) present simple → past simple to generalise about the past
  b) present simple → present perfect with the sense 'by now' or 'by some future date' rather than 'at some future date'
  c) *wouldn't die* → *might not die*, *rained* → *were to rain* for a more tentative hypothesis
5 a) other present tense verb forms: *isn't raining* (now or at some future time), *hasn't rained/hasn't been raining* (see 4b above)
  b) other past tense verb forms: *was raining* (now or at some future time), *had rained/had been raining* (before now or before some future date, especially common with a progressive form: *wouldn't be dying now*)
6 a) *may, might, could* (more tentative); *should* (= are likely to, often with a desired result)
  b) *might not, shouldn't* (more tentative); *couldn't* (wouldn't be able to)
7 open answers

3 1 c 2 g 3 b 4 h 5 e 6 d 7 a 8 f

Worksheet 5

Aim: to explore features of individual words

1 open answers

2 1 adjective (*an honest person*), adverb (*I did do it! Honest!*)
2 dis- (*a dishonest person*)
3 -ly (*I honestly believed her.*), -y (*Honesty is the best policy.*)
4 Examples: stress pattern (<u>hon</u>est), silent *h*, pronunciation of *est* (/ɪst/)
5 open answers
6 Examples:
  a) [= not telling lies] *truthful, straight, open* (person/answer)
  b) [= not cheating, stealing, or breaking the law] *incorruptible, straight, reputable, law-abiding* (person/organisation)
  c) [= not hiding your feelings: neutral] *frank, candid, direct, straightforward* (person/statement)

d) [= not hiding your feelings: potentially rude] *forthright*, *blunt* (person/statement), *outspoken* (person), *bald* (statement)

Longman Language Activator (1993) is a useful resource for teachers and students who are looking for the best way of expressing a particular meaning. Contexts of use are explained and exemplified here for words (and phrases) with the basic meanings 'honest' and 'dishonest' (see 7 below).

7 Examples: *corrupt, crooked, bent, underhand, unscrupulous, fraudulent, devious, sly, untrustworthy, sneaky, disreputable, dodgy*

8 Examples: *an honest answer/opinion/living/ man/face/broker*

9 Examples: *Honest to God . . . , To be (perfectly) honest . . . , scrupulously honest, to make an honest woman/man of you . . .*

3 1 a) incorruptible
 b) that some judges can be corrupted, e.g. bribed
 c) open answers

2 Open answers, but here is a suggestion for a). In Britain, a cup of tea is generally black tea with milk drunk from a cup (and saucer) or mug. It may well be made from a tea bag (in a mug or teapot) rather than tea leaves (in a teapot), and is drunk at any time of the day, with or without food.

4 1 a)/b) Examples:
 [with the meaning 'therefore'] I invited you, *so I'll pay.*
 ['with the purpose'] She's going to babysit *so that they can both go.* Let's check, *so as to be quite clear about the implications.*
 ['true'] *It just isn't so.*
 ['that' – referring back to something that has just been said] *I think/hope so. If so*, I'll help you.
 ['also'] *So do we.*
 ['like this'] *It looked about so long.*
 ['very'] *It was so pretty!*
 ['to such a degree'] *He was so clever that* he took his exams early.
 c) to introduce a consequence, e.g. She's ill, *so she can't come.* Similar words/phrases: *as a result, because of this, therefore.*
 to state a purpose, e.g. I need some extra work *so (that) I can pay for the holiday.* Similar words/phrases: *to (pay), in order to (pay).*

2 a)/b) The reason why/that I employed you was to improve our general efficiency. (*before*) We need to cut staff in every department. For this/that reason, I am afraid we are forced to let you go. (*after*)

5 open answers

Worksheet 6

Aim: to identify appropriate contexts for language practice

1 1 Examples: a) *will* + infinitive, *going to* + infinitive, adverbs expressing degrees of probability, first conditional
 b) (dis)agreeing (e.g. *So do I, I don't agree*), interrupting (e.g. *Yes, but . . .*)

2 Examples: a) (polite) questions, question tags, first conditional
 b) asking for/giving advice (e.g. *You could . . .*); giving/accepting thanks (e.g. *It's a pleasure.*)

3 Examples: a) prepositions of place, adverbs expressing degrees of probability
 b) making logical deductions (e.g. *It can't be . . .*), asking for/giving opinions (e.g. *Which do you think . . . ?*)

2 a) thank   b) accept/refuse   c) suggest a solution   d) give an explanation
 e) apologise/present a counter-argument
 f) reassure   g) thank   h) accept/refuse

3 1 e) Example: We used to go to Ireland for our holidays.
 2 j) Example: While she was choosing some frozen food, he put his hand in her bag.
 3 g) Example: The wiring and plumbing have been installed.
 4 h) Example: How much work have you been taking home?
 5 d) Example: She said that she enjoyed working with Woody Allen.
 6 i) Example: If I were you, I'd see a specialist.
 7 c) Example: You have to slow down when you see this sign.
 8 f) Example: I'd rather fly via Singapore.
 9 b) Example: I hope to find some holiday work.
 10 a) Example: Could you tell me how to get to the bus station?

4 open answers

5 1 Examples: eliciting personal experiences/ fantasies (e.g. Exercise 3d), a gap-fill text (e.g. Exercise 3e), a map and tourist leaflets (e.g. Exercise 3f), blackboard drawings (e.g. Exercise 3g), video without sound (e.g. Exercise 3h), a plastic spider (e.g. Exercise 3i), mime (e.g. Exercise 3j)

2 open answers

Workbook Answer Key and Tapescript

Unit one

LANGUAGE FOCUS

1 Tapescript

JAMES: When are you leaving?

JULIA: Pardon?

JAMES: When are you leaving for Germany?

JULIA: I don't know. Oh, James ... this is really difficult! I don't know what to do about this flat.

JAMES: Is it yours? Do you own it?

JULIA: No, I rent it. But all the furniture is mine.

JULIA: Hello? Speaking. Mike! Hello. I'm fine, thank you. No, I'm not working for MAP any more. I'm going to Germany – I think. Why? Really? Let me think about it. Bye.

JULIA: Well, well, well ... That was Mike Roberts of International Promotions.

JAMES: International Promotions ...

JULIA: Yes, I used to work for them. Before I moved to MAP.

JAMES: Oh yes. What did he want?

JULIA: He offered me some freelance work. He thinks I ought to start my own advertising agency.

1 1 c   2 c   3 a   4 b

2 1 T   2 F She's thinking of going to Germany
3 F She rents her flat.   4 F She used to work for International Promotions.

3 Possible answers:
1 How are you, Julia?
2 Are you still working at MAP?
3 What are you doing, then?

4 Conversation 1:
2 a) Annie to Tom
3 g) Tom to Annie
4 d) Annie to Tom

Conversation 2:
1 e) Diana to George
2 c) George to Diana
3 b) John to Diana

5 1 On Tuesday Teresa and Marco are having dinner together at a Thai restaurant in the evening.
2 On Wednesday Marco's playing badminton. Teresa's not doing anything.
3 On Thursday Teresa and Marco are both going to the camera club.

6 1 a   2 d   3 c   4 a   5 d   6 b   7 c   8 b   9 b   10 a

7 1 Where did Mary use to live? She used to live on a farm.   2 Did she use to ride to school? No, she didn't.   3 Did she use to enjoy the lessons? Yes, she did.

EXPLORING VOCABULARY

2 1 3 of: barge   houseboat   (launch)   tug
2 brick   concrete   stone
3 slate   tile
4 5 of: caravan   (cave)   cottage   (estate)   farmhouse   houseboat   tent

3 1 playing cards   2 a watering can   3 a swimming pool   4 a sleeping bag   5 running shoes

4 Tapescript/key
1 community   environment   executive   professional   responsible
2 4 of: area   caravan   industry   left-over   magical   memory   passenger   playing card   relevant

5 1 row   2 launch   3 launch/row/sink   4 launch   5 sink   6 row
The word *row* is pronounced differently depending on its meaning.

SHORT STORY

3 1 T   2 F   (She didn't talk to him about personal things.)   3 F   4 F

4 1 It means that the eyes were wide open, although he was dead.
2 His height and weight were average.
3 Letters that are sent out to thousands of people advertising products.
4 police scientists

Unit two

LANGUAGE FOCUS

1 1 b   2 c   3 a   4 b   5 a   6 b   7 b   8 a

2 1 Would you rather see ...   2 Would you like to see ...   3 Do you want to see ...

3 Tapescript

JAMES: What have you decided to do?

JULIA: Well, I've thought about it a lot, and I've decided to stay in London and form my own company ... Marsh Advertising!

JAMES: Congratulations!

JULIA: What about you?

JAMES: Me?

JULIA: Are you going to stay at MAP?

JAMES: Yes. Well, I think so.

JULIA: I see ... you'd prefer to live in the country.

JAMES: No ... of course I'd rather live in London, but I can't.

JULIA: Yes, you can. You can stay here and work with me.

JAMES: I prefer working for a bigger company.

JULIA: You haven't tried working for a small one. Look at these. My business plan for the company ... and my first job. It could be *our* first job.

JAMES: Julia ... about this new company of yours ...

JULIA: Yes?

JAMES: I think I'd like to join you.

JULIA: But James ... you prefer working for *big* companies.

JAMES: Yes ... No. I prefer working with people I like. And I'm impressed with this.

JULIA: James, I'm very pleased.

3 1 She's going to stay in London and form her own company.
2 He'd rather live in London.
3 He prefers working with people he likes.
4 He's going to join Julia in her new company.

4 1 Neither/Nor does James. 2 Neither/Nor would Julia. 3 So does James. 4 So is Julia.

5 1 So do I./I don't. 2 So did I./I didn't. 3 Neither have I./I have. 4 Nor can I./I can. 5 So am I./I'm not.

6 Possible answers:
1 It was so funny (that) I fell off the sofa.
2 It was so frightening (that) I had to leave the room.
3 It was so sad (that) I cried.
4 It was so boring (that) I almost fell asleep.
5 It was so exciting (that) I found it difficult to study.

7 1 It's like being in prison.
2 It was like flying.
3 It's like a painting.
4 They were like rocks.
5 It was like reading someone's diary.

EXPLORING VOCABULARY

2 1 nightmare 2 skyscraper 3 earthquake
4 breeze 5 whisper 6 scent 7 chess 8 sliver

3 1 flash 2 flicker 3 sparkle

4 depressing B   satisfied A
ultimate A   brilliant A
production B   seductive B
typical A   runaway A

5 1 seductive (adjective) 2 favourite (adjective)
3 breath (noun) 4 nag (verb)

6 -ment, noun, assortment, entertainment
-ly, adverb, hopelessly, happily
-tion, noun, attraction, station
-al, adjective, typical, logical

SHORT STORY

1 The mistakes are in italics:
A friend found the body of an *old* man in his *sitting room*, and called the *ambulance service*. Some *personal* letters were on the table beside the body. Inspector Temple believes that the man's death is the result of a *burglary*.

Sample corrected text:
A friend found the body of a middle-aged man in his kitchen, and called the police. Some junk mail was on the table beside the body. Inspector Temple believes that the man's death is the result of a heart attack.

3 1 To give him her report on how the man died.
2 She tells Temple that the man did not die naturally. A poison – belladonna – had been found in his blood.

5 Cause of death: poison   Murder: yes
Probable time of death: between 7 a.m. and 8 a.m.
Clues: know time of death and cause of death, letters?

Unit three

LANGUAGE FOCUS

1 1 have to 2 don't have to 3 mustn't 4 needn't
5 have to 6 have to

2 1 Must we check in now?
2 He doesn't have to hurry.
3 Does she need to change any money?
4 They needn't worry about food and drink.

3 1 Please don't smoke now./Please put your cigarette out./Please extinguish your cigarette.
The steward told her not to smoke.
2 Could you pass that magazine, please?
The woman asked him to pass her the magazine.
3 Oh, please, can I have one of those, Dad?
He begged his father to buy him a toy.
4 Get off the plane immediately!
The pilot ordered the young men to get off the plane.

4 Tapescript

OFFICIAL: Sir ... would you come here, please? That's all right, sir. You don't need to show me your passport. Where are you travelling from?

KARL: Munich, in Germany.

OFF: And what is the purpose of your visit?

KARL: Business.

OFF: I see. And what's inside the suitcase?

KARL: Oh ... normal things ... clothes ...

OFF: And, er ... what about that?

KARL: The briefcase?

OFF: Yes. Could you open it, please? And what's this – a computer?

KARL: Yes. It's a wonderful machine. My company makes them. Look ... it works like this.

OFF: That's all right, sir, you don't need to give me a demonstration.

KARL: Oh. OK.

4 1 T   2 F   3 T   4 T   5 F

5 I didn't have/need to show   2 didn't ask him to
3 answered/had to answer   4 him to   5 didn't
need/have to

EXPLORING VOCABULARY

2 1 dreadful   2 frontier   3 miserable   4 immense
5 marvellous

3 1 hang-gliding – SPORTS   2 porter – HOTEL STAFF/JOBS
3 scared – FEAR ADJECTIVES   4 vision – SENSES

4 1 confident   2 hire   3 illegally   4 relief

5 Pattern A (● ●) Each word contains the schwa
sound /ə/   Sample words: order, builder, painter

6 disappoint, disappointment   disgust, disgust   relax,
relaxation   tolerate, tolerance

SHORT STORY

1 1 Inspector Temple, Ron Elliott, Mrs McCarthy,
Detective Constable Mitchell, the doctor
2 Mr Elliott's kitchen, Inspector Temple's office

2 1 Because he knows there is a murderer who has not
been caught.
2 Because the murder of the second man seems to be
similar to that of the first.

3 1 inquest   2 verdict   3 constable   4 case

4 1 Both men lived alone.
2 Both were in their mid-thirties.
3 Both men died at the breakfast table.
4 There was no obvious cause of death in either case.

6 Tapescript
The heavy curtains were closed and the only light in the
room came from a small lamp on the table next to her old
armchair. Newspaper cuttings filled one of the walls.
SECOND MAN FOUND DEAD. BELLADONNA
MYSTERY CONTINUES. Across a photograph of Copes
was a large number two in red ink.

The woman sat in the chair and glanced lovingly at a
man's picture in a frame beneath the lamp. A tear trickled
down her face. 'Don't worry, my dear,' she whispered.
'I'll get them for you ... all of them.'

She picked up two small printed labels, turned them
over and placed them carefully on the surface of the
table. Then she reached for a small bottle, opened it
slowly, took out the dropper and squeezed four drops of
the liquid onto the back of each label. She took one of
the brown envelopes from the pile at her feet and placed
inside a folded sheet, another envelope with a printed
address, and the two labels. As she put the labels into the
envelope she looked briefly at the words on the front of
each: *I CLAIM MY PRIZE. My lucky number is 41369.* A

half-smile crossed her face and her eyes narrowed as she
sealed the envelope.

6 1 From poisoning with belladonna. They took the
poison when they licked the back of the labels to
claim their prizes.
2 Because the post had just arrived.

Unit four

LANGUAGE FOCUS

1 Tapescript
JAMES: Morning, Julia.
JULIA: Morning, James.
JAMES: Did you go to the cinema last night?
JULIA: Yes.
JAMES: Did you enjoy the film?
JULIA: Yes, it was very good. But the strangest thing
happened when I got home.
JAMES: What?
JULIA: Well, when I got home, I came up the stairs and
started to put my key in the door ...
JAMES: Yes?
JULIA: And I heard a noise. A noise like a computer – in my
flat. I thought it was a burglar.
JAMES: So what did you do?
JULIA: I ran into the street and phoned the police from the
phone box on the corner. They came very quickly. But
while we were walking up the stairs, my door opened.
I thought: Oh no! The burglar's coming out! But it
wasn't a burglar.
JAMES: Who was it?
JULIA: It was Karl!
JAMES: Karl?
JULIA: You know, Karl Schiller from Munich.
JAMES: Oh, Karl! Karl?
JULIA: Yes.
JAMES: What was he doing here?
JULIA: He was waiting for me to come home. Imagine ...
while I was sitting in the cinema, Karl was sitting in my
flat!
JAMES: How did he get in?
JULIA: The porter let him in.

1 1 arrived/got   2 went/walked   3 started/began
4 heard   5 ran   6 phoned/called/rang   7 was
going/walking   8 opened   9 was
10 was waiting   11 was (sitting)   12 was (sitting)

2 1 What was Julia doing yesterday evening?
2 Why did she phone/call/ring the police?
3 What was Karl doing in her flat?
4 What happened while the police were going up the
stairs?
5 How did Karl get into the flat?

3 1 Yes, she did.   2 No, they weren't.   3 Yes, he was.
4 No, she didn't.

4 1 When I saw my friends, I stopped to speak to them.

2 While we were having lunch, the door bell rang./We were having lunch when the doorbell rang.

3 He was watching television while she was studying./While she was studying, he was watching television.

5 1 themselves 2 yourself 3 yourselves 4 herself
5 ourselves 6 myself 7 yourself

6 1 agree 2 share 3 don't 4 think 5 disagree
6 wrong/easy

7 1 began 2 drank 3 thought 4 found 5 gave
6 shut 7 wrote Infinitive: to bring

EXPLORING VOCABULARY

2 1 burglary embezzlement mugging murder
robbery shoplifting theft vandalism
2 burglar mugger murderer pickpocket
robber shoplifter thief vandal

3 1 a detective 2 a witness 3 an enemy
4 a companion

4 1 defend 2 fiction 3 impatience 4 justice

5 Tapescript
'rebel re'bel 'import im'port 'export ex'port
'record re'cord 'increase in'crease 'decrease
de'crease

5 1 verb 2 noun
The rule for *rebel* is true for the other words.

6 running failing slipping sitting swimming
stopping shopping sleeping

HELP YOURSELF

1 ENGLISH: trunk

2 A – 6 B – 4 C – 7 D – 3 E – 2 F – 1 G – 5

3 Tapescript
A I'm looking for that white stuff... that white powder with a lovely smell. You know, you put it on your body after you have a bath. What's it called?
B I need to borrow a thing ... a metal thing for making holes in the wall. It's got a handle, and you put long pieces of metal in the other end, different sizes, and it works with electricity. What do you call that?
C I want to buy that stuff for ... for putting on your hair when you wash it. It's a liquid, and you buy it in bottles. No, not shampoo. It makes your hair soft and shiny. Do you know what I mean?

3 A ENGLISH: talcum powder B ENGLISH: electric drill
C ENGLISH: conditioner
Uncountable: liquid, stuff, powder
Countable: object, thing

4 Possible answers:
1 It's a strange shape: there's a handle that you hold and a long round part that spins. The handle is usually made of metal, and the other part is made of a soft material. It can be any colour (but the soft part is often yellow or white). It's for painting with; it's like a paintbrush, but you can paint much faster with it.
2 It's a large round object, like a bowl with holes in it and a long handle. It's made of plastic or metal. It can be any colour. It's for draining vegetables after you cook them, but you can also use it to wash salad and fruit. It's like a sieve, but the holes are bigger.

Unit five

LANGUAGE FOCUS

1 1 Are (the) advisors paid?
2 Where are they sent?
3 What is each advisor asked to do?
4 How are the businesses helped?
5 Are the advisors trained for the job?
6 How is BESO supported?

2 1 Computers for children are designed by Karl's company.
2 Computer software is also developed by his company.
3 The computers are advertised in Britain.
4 Three important features are included.
5 Children are asked to read the screen.
6 Their answers are spoken.
7 Julia and James are offered the job of advertising the product.

3 Tapescript
JAMES: So, Karl, why are you here?
KARL: I'm here to talk about a new business project.
JAMES: What kind of project?
KARL: We are developing a new computer for children to use at home.
JULIA: But surely lots of computers are made for children. And several of them are made for home use. How is your computer different?
KARL: In two ways. First, we are developing a better voice-activation system.
JAMES: A what? What does 'voice-activation' mean?
KARL: I'm sorry. Look at this computer. This one doesn't have a voice-activation system. In order to use it, you have to read the screen, then type your answer. But the new computer speaks to the child. It asks a question and the child answers by speaking to the computer.
JULIA: Hmm. And the other difference?
KARL: Ah. It's educational. Nearly all the computers that are sold to children are for games. Hardly any are made to help children with their studies.
JAMES: But yours will.
KARL: Yes, because we're developing special software – study programs.

JULIA: I see. You're developing the software in order to sell the hardware.

KARL: Yes, you could say that.

JULIA: Hmm. It sounds very interesting.

KARL: It is. And maybe you two can help me.

JULIA: How?

KARL: We need an advertising agency, here in Britain.

JAMES: You mean you want us to advertise your computer?

KARL: Yes.

JAMES: You realise that we're a very small company.

KARL: Yes. So you must work for me to get bigger. What do you think?

JULIA: Yes, Karl. We'd love to.

3 1 T   2 T   3 F   4 F   5 F   6 T   7 T

4 1 all of them   2 none of them   3 neither of them
4 both of them

5 1 c   2 b   3 d   4 a   5 b   6 c   7 a   8 d

EXPLORING VOCABULARY

2 1 distribute   2 design   3 peel   4 employ

3 1 C   2 S   3 S   4 C   5 S/C   6 C   7 S

4 1 fuel tank   2 book-keeping   3 hand-made
4 water tank   5 note-taking   6 home-made

5 advertise   en<u>cou</u>rage   em<u>plo</u>y   re<u>pay</u>
Employ and *repay* have the same pattern.
ad<u>ver</u>tisement   em<u>ploy</u>ment   en<u>cou</u>ragement
re<u>pay</u>ment   ad<u>ver</u>tisement (<u>ad</u>vertise)

HELP YOURSELF

2 A 5   B 8   C 3   D 6   E 1   F 2   G 7   H 4

3 1 Good evening.   2 Sample answer: Thanks very much.   3 Sample answer: By credit card.   4 Cheers!

Unit six

LANGUAGE FOCUS

1 Possible answers:
1 If I lived in a house like that, I'd have a motorbike to drive around it!
2 If I visited Kauai, I'd try to go when it's not raining!
3 If I spoke 58 languages, I'd get a job as a translator.
4 If I received 33 million cards, I'd put an advertisement in the paper to thank people.
5 If I walked that far, I think I'd want to relax for a few years!
6 If I earned that much money, I'd give most of it away.

2 1 What will she do if it rains? She'll go home.
2 How will/would she feel if she sells/sold everything? She'll/She'd be delighted!
3 How would she react if the police came? She'd run!
4 What would happen if the police caught her? She'd have to pay a fine.

3 1 B: I'm terrified of spiders.
    A: Don't be afraid.
2 B: I'm worried about (failing) my driving test.
    A: Don't worry. It'll be all right.
3 A: Calm down. Don't cry.

4 Tapescript
JAMES: Good morning, Julia.
JULIA: Hello, James.
KARL: Good morning, James.
JAMES: Karl! What are you doing here? Aren't you going back to Germany today?
JULIA: He isn't very well.
JAMES: What's the matter with you?
KARL: I don't know. I have a . . . er . . .
JULIA: A sore throat.
KARL: Yes, I have a sore throat. And a headache. I feel awful.
JAMES: Have you got a temperature?
KARL: What do you mean?
JAMES: Are you very hot? Do you have a high temperature?
KARL: Oh, yes. I think so.
JULIA: I think you've got 'flu.
JULIA: You ought to see a doctor as soon as possible.
KARL: Here in England? I don't know.
JAMES: I think you should go back to your hotel, and you should call a doctor.
KARL: But I have to be back in Germany tomorrow.
JULIA: Karl, you're ill. I think you ought to stay here for a few days. What do you think, James?
JAMES: No, Julia, he must do what he wants to do. Karl, if I were you, I'd go home tomorrow.
KARL: Listen . . . it's all right. I think I should go back to Germany and have a rest for a couple of days. Then I'll be fine. Really.
JULIA: OK, if that's what you want to do.
KARL: Thank you for all your help. Both of you. You've been very kind.
JULIA: Don't mention it. I hope you feel better soon.
KARL: Thank you. I'll call you next week about advertising the new computer.
JULIA: Good.
JAMES: OK. I'll show you out.
KARL: Thank you. Bye bye.

4 1 Today.   2 He's ill.   3 Sample answer: He should see a doctor.   4 . . . call a doctor to his hotel and then return to Germany.   5 . . . stay in Britain for a few days.   6 . . . go back to Germany and rest for a few days.

5 1 He ought to be more polite.
2 He shouldn't use the phone for/make personal calls.
3 He oughtn't to miss appointments.
4 He should ask when he wants to take time off.

EXPLORING VOCABULARY

2 1 pharmacist   2 optician   3 surgeon   4 dentist

5 herbalist   6 acupuncturist

3 1 creak   2 growl   3 drip   4 rustle   5 moan/scream
6 splash

4 1 anxious   2 afraid   3 petrified   4 irritable
5 overweight   6 sore

5 medicine   psychiatrist   rustle   honest   knee
lamb   wrong   could   answer

6 1 adjective – healthy   2 verb – be the right size
3 verb – put in place   4 noun – sudden attack
5 verb – match

SHORT STORY

1 Possible answer:
She's about 25 and she's got long dark hair. She's
wearing summer clothes and she's got a camera over
her shoulder.

2 1 On a Mediterranean beach.
2 No, she's on holiday.
3 It's probably summer.

3 1 T   2 F   3 T   4 F   5 F

4 It's probably a romance.

Unit seven

LANGUAGE FOCUS

1 1 which/that/–   2 who/that   3 who/that/–
4 who/ that   5 that/which/–   6 that/which/–
7 that/which   8 who/that   9 who/that
10 that/which   11 that/which/–   12 that/which
13 that/which/ –   14 that/which/–

2 1 We visit the cottages that/which Mr James owns.
2 Betty is one of the cleaners who/that works for
Mr James.
3 She does a lot of extra work that/which visitors
create for her.

3 Possible answers:
1 Mr and Mrs James often need to have the curtains
cleaned.
2 They sometimes need to have broken plates
replaced.
3 They have the outsides of the cottages repainted
every year.
4 They have the carpets shampooed.
5 Sometimes they need to have pieces of furniture
repaired.
6 They have the rooms redecorated every winter.

4 1 A: Excuse me. Can I have this dress cleaned, please?
B: Of course. Would you mind waiting a moment?
2 A: Can I have this ice bucket filled?
B: Of course. Would you mind asking the barman?

5 Tapescript and key
Answers to exercise in *italics*.

Part one.
RECEPTIONIST: Good afternoon, sir.
JAMES: Good afternoon.
REC: Terrible weather.
JAMES: Yes.
REC: Excuse me, but … what happened to you?
JAMES: A car went past and splashed me … My name is
Brady, by the way. James Brady. I have a reservation.
REC: Oh yes, Mr Brady.
Part two.
REC: You're the person who asked about a conference
room.
JAMES: *That's right. (1)*
REC: Well, our small meeting room is free tomorrow.
JAMES: *Excellent. That's just what I wanted. (2)*
REC: Good. And you're staying for two nights.
JAMES: *Yes. (3)*
REC: Bed and breakfast only.
JAMES: *That's right. (4)*
REC: Would you mind filling in this form, please?
JAMES: *No, not at all. (5)*

5 A car splashed him.
See tapescript above for answers to second part.
I'll take it. 2
Certainly. 5
Yes, I am. 1, 3

EXPLORING VOCABULARY

2 1 annual   2 bewildered   3 proud   4 rapid

3 A swarm   B flick   C roll   D scatter   E leap   F hurl

4 cheap /tʃ/   school /k/   character /k/   lunch /tʃ/
architecture /k/   chaos /k/   drench /tʃ/

5 a) cleaning done without water
b) catering done by yourself
c) to put something back
d) cleaning done in spring-time
e) you serve yourself
f) to pay back

SHORT STORY

1 Elaine, Toni and Jenny. The others are Toni's friends.

3 1 c   2 a   3 c   4 b   5 a

4 1 They share an interest in literature and writing. Toni
is a student while Elaine is a journalist. She lives alone
and rarely sees her parents. He lives with his parents.

Unit eight

LANGUAGE FOCUS

1 1 a   2 b   3 a   4 a   5 b   6 a   7 d   8 c   9 c   10 d

2

INFINITIVE	PAST TENSE	PAST PARTICLIPLE
blow	blew	blown
creep	crept	crept
hide	hid	hidden
shoot	shot	shot
sink	sank	sunk
spill	spilt	spilt
spring	sprang	sprung
steal	stole	stolen

3 Tapescript

JAMES: Paul . . . Cook?

PAUL: Yes, sorry . . . I don't . . .

JAMES: James Brady.

PAUL: Of course. How are you?

JAMES: I'm fine, thanks.

PAUL: Would you like to join me?

JAMES: Thank you very much.

PAUL: Are you still with MAP Advertising?

JAMES: No, I have my own company now. Well, I'm working with Julia Marsh.

PAUL: Oh, I remember Julia. How is she?

JAMES: Fine.

PAUL: How long have you been working together?

JAMES: For about three months . . . since May. What about you, are you still working for . . . um . . .

PAUL: Art and Design.

JAMES: Art and Design.

PAUL: Yes, I've been working there for ten years.

JAMES: What are you doing here in Manchester?

PAUL: I have an interview for a job.

JAMES: A job?

PAUL: Yes, I've been looking for a new job for ages. I've applied for a job with a newspaper company, here in Manchester.

JAMES: What kind of job?

PAUL: Well, they're starting a new weekly magazine for young people, and they're looking for a designer.

JAMES: Have you worked on a magazine before?

PAUL: No, but it's always been my ambition. Oh, I must go. I don't want to be late.

JAMES: Well, good luck.

PAUL: Thank you.

3 1 How long has James been working in the new company with Julia?

2 How long has Paul been working at Art and Design?

3 How long has Paul been looking for a new job?

4 Sample answer: What kind of job has Paul applied for?

5 What has Paul's ambition always been?

4 1 Sorry, I mean Bob.

2 Pardon?

3 I'm afraid I don't understand.

4 Could you give me an example?

5 Could you say that again?

EXPLORING VOCABULARY

2 1 literature 2 geography 3 biology 4 economics 5 architecture 6 physics

3 1 engineer 2 journalist 3 chef 4 solicitor/lawyer 5 veterinary surgeon (vet) 6 plumber 7 librarian

4 1 legal 2 patient 3 complex 4 tolerant 5 renewable

5 Tapescript and key

biology anthropology technology ecology geology astrology sociology psychology zoology   The ending -ology means 'the science/study of something'.

SHORT STORY

All exercises have open answers.

2 Tapescript

Elaine sat on the balcony of their hotel room with a glass of fresh orange juice in her hand and gazed down at the peaceful scene below. Verino, a town that she had come to love, was beautiful in the clear early morning light. But she was thoughtful and sad. The holiday was almost at an end and, try as she might, she couldn't get her feelings for Toni out of her mind. As far as she could tell, though, Toni himself accepted without question that their friendship would soon be over.

'What shall I do, Jenny?' Elaine asked her friend, when she came out to join her for breakfast. 'I've never met anyone that I like so much. He's kind, he's interesting . . . but he lives so far away.'

'Oh, come on, Elaine – be realistic. You're on holiday here so you're happy and relaxed, and he's given you some attention. But that's all it is – a holiday relationship. You love your job – do you want to give it up? What would you do? And why on earth would he want to come to England? He's never talked about the possibility, has he?'

'No, no, he hasn't. And – well, I've got no idea how he really feels about me. I know I'm being silly.'

'You probably are – but it's your life. Why don't you just talk to him? Tell him how you feel and see what he says. Then at least you'll know if you've got any options to choose from.'

That evening was the last one that Elaine and Toni had together, and they ate together in a quiet restaurant just outside the town. At the end of the meal, Elaine took a deep breath.

'Toni,' she said slowly. 'I'm going to miss you, I really am. I don't know how to tell you, but . . .'

Toni reached across the table and put his hand over hers. 'You don't need to tell me, Elaine. I feel the same. But I haven't spoken about it because it's impossible. I can't come to England – I have my work and my studies . . . and when I finish my studies, I must help my parents in the business. I want to be here, and I would like very much for you to be here too, but what can we do?'

'Well, I've got to go tomorrow, you know that . . . but if I *can* find a way to come back – after all, I'm a journalist, and they need reporters overseas – if I *can* find a way, will you truly be happy to see me?'

'You know I will, Elaine. Write to me for now and depend on it, I'll wait for you.'

Unit nine

LANGUAGE FOCUS

1 1 That looks delicious!
2 He looks like a soldier.
3 He looks like his father.
4 They still don't look clean!
5 He doesn't look ill!
6 They don't look like sheep.
7 That film doesn't look very funny.

2 1 C   2 B/D   3 F   4 E   5 A

3 Possible answers:
1 I'm not sure.   2 Yes, maybe.   3 No, I'm sure they won't.   4 No, they definitely won't.   5 Yes, almost certainly.

4 Tapescript
JULIA: James, you know that model agency you went to see – when you were in Manchester?
JAMES: Yes?
JULIA: What did you tell them?
JAMES: I asked them to send us photographs of children – from 12 to 14 years old – that we could use to advertise an educational computer.
JULIA: Well, they sent photographs. But most of these aren't children, are they? I mean, look at this one.
JAMES: What's the matter with him?
JULIA: Well, he's too old. He must be at least 18. He could be 23.
JAMES: I don't think so.
JULIA: He must be! He looks like a professional footballer.
JAMES: It says here he's 16.
JULIA: He can't be! It must be a mistake. And look at this one.
JAMES: She looks great!
JULIA: Yes, she looks great, but she doesn't look like a schoolgirl, does she?
JAMES: I don't know . . . schoolgirls look like that these days.
JULIA: She doesn't look like a schoolgirl who enjoys using a computer. She looks like a rock singer. And it says she's 14! She can't possibly be 14.
JAMES: What about this one? She looks about 14, she looks intelligent, and she looks like someone who enjoys doing homework.
JULIA: James . . .
James: What?

JULIA: That's a photograph of me.
JAMES: Oh.
JULIA: When I was at university.

4 1 He must be at least 18. He could be 23.
2 He looks like a professional footballer.
3 He's 16.
4 She doesn't look like a schoolgirl.
5 She can't possibly be 14.
6 She looks 14. She looks intelligent. She looks like someone who enjoys doing homework.
7 About 18–21!
8 Because the models from the agency look too old and don't look serious enough. The other one is of Julia!

EXPLORING VOCABULARY

2 A shave   B tattoo   C dye   D pierce

3 1 unshaven   2 smart   3 slim   4 drab/dull
5 unconventional
a) clean-shaven   b) slim   c) conventional
d) drab, smart

4 1 too crowded   2 too hot   3 too full   4 too busy

5 <u>colour</u>ful (1)   ag<u>gres</u>sive (1)   <u>inn</u>ocent (2)
con<u>ven</u>tional (3)   re<u>bell</u>ious (1)   for<u>get</u>ful (2)
sym<u>met</u>rical (1)   un<u>shaven</u> (1)
The /ə/ sound never occurs in a stressed syllable. These adjective endings commonly contain the /ə/ sound: -ent, al, -ous, -en. While the ending -ful is usually transcribed /fʊl/, it can also be interpreted as /fəl/ if it is said quickly.

HELP YOURSELF

1 1 stationary   2 stationery   3 I need to buy some stationery.   4 same

2 1 a) Boxing Day b) boxer c) box d) Boxing e) the box
2 boxcar   3 the box (meaning TV), box someone's ears   4 four   5 in reply to newspaper advertisements
6 buy tickets   7 never

3 1 appearance, foreigner, necessary, psychology
2 foreigner, necessary

Unit ten

LANGUAGE FOCUS

1 1 managed   2 couldn't   3 managed/was able
4 wasn't able   5 can   6 managed   7 will be able

2 1 Didn't they manage to book seats?
2 Could he speak Spanish before he moved to Mexico?
3 Will you be able to leave at the same time as me tomorrow?
4 Is she able to learn her lines?
5 Can you help me, please?

3 1 D   2 G   3 E   4 F   5 A   6 C   7 B

4 Tapescript
JAMES: Julia, I'm so sorry! I couldn't find a taxi.
JULIA: So how did you get here?
JAMES: By taxi.
JULIA: I thought you said you couldn't find one.
JAMES: I meant I wasn't able to find a taxi when I came
out of my flat. What's the matter with you?
JULIA: Nothing. Here's your ticket.
JAMES: Thank you. How much did it cost?
JULIA: £27.
JAMES: £27? For a jazz concert?
JULIA: This is The Wilton Hall. It's expensive.
JAMES: But weren't there any cheaper tickets?
JULIA: Yes. But I wasn't able to get any.
JAMES: Why not?
JULIA: Because I forgot to get them last night ... Come on,
we'll miss the beginning.

4 1 Because he couldn't find a taxi.
2 He managed to find one after some time.
3 They cost £27 each.
4 It's called The Wilton Hall.
5 Julia wasn't able to get cheaper ones.

5 1 I did.   2 I wasn't.   3 she wasn't.   4 I won't.
5 we could.   6 I can't.

6 1 I am writing   2 In my opinion   3 as far as I'm
concerned   4 Why don't   5 certainly   6 Let's

EXPLORING VOCABULARY

2 1 dolphin   2 bear   3 camel   4 tiger   5 monkey
6 seal   7 crocodile   8 elephant   9 lion

3 A feed   B operate   C raise   D attach

4 1 ringmaster   2 clowns   3 acrobat   4 cages
5 tricks   6 box office

5 greed   sophistication   sensitivity   laziness
honesty   tenderness

HELP YOURSELF

1 Tapescript
A Can I have some clean 'towels, please?
B Can I have some 'clean towels, please?

1 1 B clean   2 A towels

2 Tapescript
1 Have you got any 'fresh 'fish?
2 Have you got any 'fresh fish?
3 Have 'you got any fresh fish?

4 Do you like 'this coat?
5 Do you like this 'coat?
6 Do you 'like this coat?
7 Do 'you like this coat?

2 1 B fresh fish   2 A fresh   3 C you   4 E this
5 G coat   6 D like   7 F you

3 Tapescript
A: Why didn't you ring me?
B: But I 'did ring you.
A: I'm a complete failure.
B: 'No, you 'aren't.

1 Why weren't you there?
2 Why isn't it Friday?
3 My hair looks awful.
4 I've never met her.

PRESENTER: Now listen to the complete dialogues.
1 A: Why weren't you there?
   B: But I 'was there.
2 A: Why isn't it Friday?
   B: But it 'is Friday.
3 A: My hair looks awful.
   B: 'No, it 'doesn't.
4 A: I've never met her.
   B: 'Yes, you 'have.

3 But I did ring you.
No, you aren't.
1 But I was there.
2 But it is Friday.
3 No, it doesn't.
4 Yes, you have.

4 Tapescript
1 MAN: It's a lovely day, isn't it?
   WOMAN: Yes, beautiful.

2 Man: It's a lovely day, isn't it?
   Woman: Yes, beautiful.

PRESENTER: Now listen and repeat.
A: It was a great lesson, wasn't it?
B: Yes, very interesting.

A: It was a great lesson, wasn't it?
B: Yes, very interesting.

4 1 a) 1   b) 2

5 Tapescript
1 Oh yes? He will? (questioning)
2 Oh yes. He will. (reassuring)
3 Oh yes he will. (expressing anger)
4 Oh yes he will. (showing fear)
5 Oh yes! He will! (expressing excitement)

5 A 4   B 2   C 3   D 1   E 5

Unit eleven

LANGUAGE FOCUS

1 1 –   2 the   3 a   4 a   5 the   6 The   7 the
8 –   9 –   10 the   11 the   12 the   13 The   14 –
15 –   16 The   17 –   18 –   19 the   20 a

2 1 Is the system likely to be expensive?
2 Are you certain the system will save money?
3 Are you seriously going to consider installing it?

4 Do you intend to think about other options?

5 Will the system work in your offices?

3 1 We're not/We aren't going to have a long discussion.

2 They won't be able to express their opinions.

3 I may not have to leave early.

4 The directors probably won't join us.

5 I'm not likely/unlikely to lose my temper.

4 Possible answers:

1 They probably won't be very happy.

2 She might fall.

3 He'll definitely go to prison.

4 They're certain to get wet.

5 I'm sure the dog won't catch her.

5 Tapescript

KARL: Julia, this is Michael Preston.

JULIA: How do you do?

MICHAEL: How do you do?

KARL: Michael Preston, James Brady.

JAMES: How do you do?

MICHAEL: How do you do? Please sit down. Now, I'm thinking of investing in the new computer that Karl's company has produced. And I asked him if I could come to this meeting to talk about the advertising campaign. I'm very glad that you could come.

JAMES: Thank you.

MICHAEL: I've read your campaign plan. It looks very good. I only have a few questions, mainly about where you intend to place these advertisements.

JULIA: Well, as you can see from the plan, we intend to place them in family magazines and on early evening television. We want whole families – parents *and* children to see them.

MICHAEL: Hm. But are you sure they'll be the right families?

JULIA: The right families? Well ...

MICHAEL: Are you certain that the parents who read those magazines and watch those television programmes are the people who will buy an educational computer for their children?

JULIA: Yes, we are quite certain. This research shows that ...

MICHAEL: Good. I'm sure you're right. And I see that you're thinking of placing advertisements in teenage magazines.

JULIA: Yes, we intend to place them in *some* teenage magazines – the more serious ones.

JAMES: And we're thinking of putting them in Sunday newspapers – because it's likely that the whole family will be together on Sundays.

MICHAEL: I see. But do you really think that parents *and* children will like the same advertisements?

JAMES: We think they'll like our advertisements.

MICHAEL: Good.

5 1 might   2 are planning to   3 may   4 probably

5 likely to

6 1 Just a minute   2 What do you mean?

3 for instance   4 Could I ask a question?

EXPLORING VOCABULARY

2 1 coal   2 gas   3 petrol (AmE gas)   4 solar power

3 1 profit – commercial (BUSINESS)

2 household – domestic (HOME)

3 politician – election (GOVERNMENT)

4 ecosystem – rainforest (THE ENVIRONMENT)

5 planet – spaceship (SPACE TRAVEL)

6 ocean – wave (THE SEA)

4 pessimist – pess<u>i</u>mistic   <u>opt</u>imist – opti<u>mi</u>stic benefit – bene<u>fi</u>cial   en<u>vi</u>ronment – environ<u>men</u>tal <u>theo</u>ry – theo<u>re</u>tical   <u>comm</u>erce – com<u>mer</u>cial gene – ge<u>ne</u>tic   <u>ri</u>dicule – ri<u>di</u>culous

5 1 a   2 an   3 –   4 –   5 an   6 a crew [no code = countable]   petrol [U]   optimist [no code = countable]   emotion [C,U]   ridicule [U] species [no code = countable]

SHORT STORY

1 Possible answer:

A boy is standing in a school yard with some older and taller boys standing around him. One of the bigger boys is smiling.

2 Possible answers:

1 T (They move because she changes jobs.)

2 T (She is probably older because she gives him advice.)   3 F   4 T   5 F (When they came up to him he thought they were being friendly at first.)   6 T

3 Because people look back at them as a time when they had few responsibilities and their worries were small.

Unit twelve

LANGUAGE FOCUS

1 Passive forms: was founded, was priced, was reduced was sold, was changed, is based, is owned, are taken, is aimed

2 1 When was *The Manchester Guardian* founded? It was founded in 1821.

2 When was the price reduced? The price was reduced in 1855.

3 Who became the editor in 1871? The writer C P Scott became the editor.

4 When was the title changed? The title was changed in 1959.

5 Where is the editor based now? The editor is based in London.

6 How often does the international paper appear? It appears every week.

7 Where is it sold? It is sold outside Britain.

3 1 …in Belfast has been bombed.
2 …to cut awards for crime victims have been delayed.
3 …have been stolen from the Prince's apartment.
4 …have been discovered by a farmer in his fields.
5 …has been sold by a collector for one million pounds.

4 Tapescript
JULIA: Here it is. 'Listen and learn with the new talking computer'.
JAMES: Hey! That looks great!
JULIA: Are you surprised?
JAMES: No! I knew it would look good.
JULIA: Now … let's look through all these newspapers and magazines.
JAMES: Why?
JULIA: To see if there are any new advertisements for other computers.
JAMES: Oh. Right.
Part two.
JAMES: Hey! There's a story here you may wish to read.
JULIA: I'm too busy!
JAMES: It's about MAP.
JULIA: Really? MAP? Let me have a look. 'The prize for the best advertising campaign of the year has been presented to MAP Advertising.
JAMES: What for? Which campaign?
JULIA: Um … hold on … 'MAP were given the prize for' … where is it … ah … 'the Drake bicycles campaign'.
JAMES: What?
JULIA: Apparently Tom Hall was presented with the prize in London last night.
JAMES: Last night? Tom was in London last night?
JULIA: Yes. I'm surprised he didn't come and see us.
JAMES: Are you surprised? I'm not.
JULIA: Why?
JAMES: Because *he* has been given a prize for something that was done by *me*.
JULIA: What do you mean?
JAMES: Drake bicycles was my account. I managed that campaign.
JULIA: Oh, I see.
JAMES: I'd like to have a word with Tom Hall.

4 1 Because their advertisement looks great.
2 Because Tom Hall has been given a prize for some work that James did himself.

5 1 has been presented   2 were given
3 was presented   4 has been given, was done

EXPLORING VOCABULARY

2 1 huge   2 silent   3 filthy   4 feeble   5 evil
6 vicious

3 1 attacked   2 injured   3 sacked   4 arrested

4 powerless   defenceless
The ending *-less* means *without*.

5 1 She's left-wing   2 a left-wing newspaper   3 He's right-wing   4 a right-wing newspaper   5 He's Chinese.   6 Chinese food.   7 She's Japanese.
8 Japanese food.

6 edit – editor   report – reporter   publish – publisher
print – printer   own – owner

SHORT STORY

1 Possible answer:
Michael has been bullied at his new school. He has told his sister, but not his mother, about the problem.

3 Bullies are people who use their strength to hurt or frighten people who are not as strong as they are.

4 1 M   2 TB   3 M   4 G   5 TB   6 T

Unit thirteen

LANGUAGE FOCUS

1 Tapescript
JAMES: I think that's Ireland. Who's the photographer?
JULIA: … 126 .. er … John Patrick Brady.
JAMES: What! That's amazing!
JULIA: Do you know him?
JAMES: He's my cousin! Here's another one. Aren't they wonderful?
JULIA: They're very nice.
JAMES: No, what do you really think? Isn't he a great photographer?
JULIA: He's good. Yes, he's very good. Is he a professional photographer, or is this his hobby?
JAMES: I don't really know. He was still a student when I last saw him. He'd always been a keen photographer – like me. But he certainly hadn't had an exhibition when I left Ireland, I'm sure of that. He'd never even *sold* a photograph, as far as I can remember.
JULIA: Why didn't he tell you about this exhibition?
JAMES: Well, I haven't seen him for ages. I'm so pleased he's doing well.

1 1 Yes, he is.   2 Yes, she does.   3 No, he hasn't.
4 No, he didn't.   5 Yes, he had.   6 No, he hadn't.
7 No, he hadn't.

2 1/2 had just brought   3 had asked   4 As soon as
5 had had   6 had called   7 had identified   8 had disappeared   9 had been   10 had noticed
11 until then   12 By the time   13 had worked

3 1 – C   2 – G   3 – B   4 – A   5 – H   6 – D   7 – F
8 – E

4 1 Go up the escalator to the second floor. Turn left and left again and you'll see the hairdresser's in front of you near the restaurant.
2 Go up the stairs to the second floor. Then turn left. Go past the hairdresser's and the restaurant is in front of you.

EXPLORING VOCABULARY

2 1 sight   2 exhibit   3 curator   4 mouldy   5 wrinkles

3 1 embarrassed   2 cheerful   3 neglected   4 genuine
5 offensive

4 <u>arrogant</u> as<u>sis</u>tance
apple (A)   attractive (B)   alive (B)   attack (B)
addict (A)   arrest (B)

5 carefree
1 crime-free   2 meat-free   3 tax-free   4 disease-free

SHORT STORY

1 1 a week   2 unhappy   3 sister   4 identify

2 Possible answers:
1 He could tell his mother.
2 He could identify the bullies to the teacher.
3 He could hit the bullies back.

3 TAPESCRIPT
Option A:
Michael agreed to go with his teacher to see the head of the school. He told them both what had happened and named the three boys who were involved. A boy who witnessed the second attack was able to confirm Michael's story, and the head sent for the three boys.
Option B:
Michael was too frightened to tell the teacher about the bullies. He knew what would happen if he did, because he had tried it before in another school. Next time he saw the boys he offered to give them his pocket money and his bus money every week if they left him alone. They agreed.
Option C:
Michael refused to talk about the attacks. He was so frightened that he decided to stop going to school. He didn't want his mother or sister to know, so he left home every day at the usual time and spent his days wandering around town. At least he felt safe.

3 Possible answers:
1 B (or none of them)   2 B   3 A

5 Tapescript
Michael didn't feel able to talk about the bullying to anyone. His mother would worry, he knew that. His sister had her own life and had never mentioned their conversation again. And he was afraid of what the bullies might do to him if he told any of the teachers. He should be able to deal with the situation on his own, he thought – and if he couldn't, well, that was his problem.

He didn't even enjoy his lessons any more, because it was a real struggle to concentrate. As he sat in class each day, he thought about what might happen after school and his mind went completely blank. And sure enough, two or three times a week, the bullies were out there waiting for him.

'Well, now, Jenkins, what've you got for us today?' they shouted. He took the money from his pockets and gave it to them without a word. Sometimes they still hit him, for the fun of it, but usually they ran off laughing.

The day everything changed was the day before half-term. Michael had stayed late at school because he needed some advice about an English project. The school grounds were empty by the time he left, but he stayed tense and watchful on the short walk to the bus stop.

There was only one other boy from the school at the bus stop, and Michael stared at him in disbelief. He was in the lowest class, so he was probably eleven years old, but he looked younger. His clothes were dirty and torn, and he was crying quietly.

'Hey, what happened to you? Are you OK?' Michael asked, but he had a sick feeling in his stomach.

'These boys ... they said they'd hurt me if ... if I didn't give them money,' the younger boy said. 'And I haven't *got* any money – only my bus pass ...'

'Were there three of them?' Michael asked quickly. 'They're from school, aren't they?'

The child looked at him with suprise and nodded. Tears dripped from his cheeks to the ground. Michael took a deep breath.

'It happens to me too,' he said, 'and there's only one way to stop it. We've got to tell someone. Come with me.'

He led the way back to the school, and found his English teacher. As Michael told his story, the boys could see sympathy and anger in her face. When he finished, there was a short silence.

'Is this right, Ben?' she said to the younger boy. 'Are you sure the bullies who hurt you are the same ones?'

'Yes,' he whispered. 'The same ones.'

She looked at each of them in turn. 'You can leave it to me now,' she said quietly. 'I'll see the head first thing in the morning. And don't worry, either of you. This won't happen again, I promise you.'

5 1 talk   2 concentrate   3 went   4 continued
5 hit   6 stayed   7 was threatened   8 realised
9 tell

Unit fourteen

LANGUAGE FOCUS

1 Tapescript
JULIA: What did you say to Tom?
JAMES: I said I was very unhappy that he hadn't spoken to me about the prize.
JULIA: And what did he say?
JAMES: Well, he said he was sorry that I was unhappy, but he also said that the *agency* had won the prize, and not an individual employee.

JULIA: Well, he's right, of course.

JAMES: I suppose so ... Anyway, he said he wanted us to be friends. And he asked me if I would go and see him ... spend a weekend in the country.

JULIA: And what did you say?

JAMES: I said I would. And I asked if you could come with me.

JULIA: Oh, did you!

JAMES: Wouldn't you like a weekend in the country?

JULIA: Hmm. I'll think about it.

JAMES: Oh, come on – it'll be fun.

JULIA: I said I'd think about it.

1 1 Yes, they are.
2 JAMES: I'm very unhappy that you haven't spoken/didn't speak to me about the prize.
TOM: I'm sorry that you're unhappy. But the agency (has) won the prize, not an individual employee. I want us to be friends, James. Would you come and see me ... spend a weekend in the country?
JAMES: Yes, I'd like that. Can Julia come with me?

2 1 d　2 b　3 a　4 b　5 b　6 a　7 c　8 d　9 b
10 a

3 1 ... me (us) there.　2 ... is ours (mine)!
3 ... then ... in a week's time (the following week).
4 ... taken him ...

4 Possible answer:
SECRETARY: Bradford College. Can I help you?
SECRETARY: No, this is the forestry department. Who's speaking, please?
SECRETARY: Hold the line a moment. I'll try to put you through to him.
SECRETARY: I'm afraid he's engaged at the moment. Would you like to call again later?

EXPLORING VOCABULARY

2 1 database, program, software
2 hold on, put through
3 a) fortnight　b) tool　c) enthusiasm
4 a) casual　b) engaged/restricted　c) familiar
d) valuable

3 1 survey　2 confirms/confirmed/has confirmed
3 average　4 cost　5 bargain　6 alarmed
7 revolution

4 unconscious　insecurity　unenthusiastic　unfamiliar
unreasonably　inconclusive　unrestricted
insignificant

5 access A　alarmed B　aware B　confirm B
consult B　danger A　delay B　dial A　drawback A
engaged B　excuse B　express B　imply B

HELP YOURSELF

This page contains only linguistic contrasts: open answers.

Unit fifteen

LANGUAGE FOCUS

1 Tapescript

JAMES: I don't know which platform we want, do you?

JULIA: It's platform 2 – I think ... but I'm not sure if it's the same at weekends.

JAMES: Excuse me?

EMPLOYEE: Yes?

JAMES: Do you know where the Banbury train leaves from?

EMPLOYEE: Yes, Platform 1.

JAMES: Is Platform 1 on this side?

EMPLOYEE: No, it's the last platform – over there.

JAMES: Oh ... Er ... Excuse me ... Could you help me, please?

EMPLOYEE: Of course. But I must deliver this parcel first. Would you mind waiting a moment?

JAMES: No, not at all ... Thank you.

JAMES: What are you reading?

JULIA: A book about India.

JAMES: Oh, I've always wanted to go there. I read a wonderful book about India when I was a boy. I don't know who wrote it. It was called *The Forgotten Land*. What's that one about?

JULIA: The writer describes a journey through India. He travelled by slow train, by boat ... he even went 200 kilometres by taxi.

JAMES: That sounds wonderful. I'd love to do something like that.

JULIA: I wonder if I would like it ...

JAMES: I think you would. Maybe we could go there together.

JULIA: Together? You and me on a slow train through India? I'm not a very good travelling companion.

JAMES: I'm sure you'd love it.

JULIA: Would I, James?

JAMES: Yes, you would.

JULIA: India ... with you ... I wonder.

1 1 ... which platform we want　2 ... if it's the same
3 ... where the Banbury train leaves　4 ... who wrote
5 ... if I would like

2 1 Which platform do we want?　2 Is it the same at weekends?　3 Where does the Banbury train leave from?　4 Who wrote it?　5 Would I like it?

3 Possible answers:
1 Can you tell me if you are open on Sundays?
2 Can you tell me what your fax number is?
3 Do you know how much a return ticket to Oxford is?
4 Do you know if the London train is on time?
5 Can you tell me if the train stops at Reading?

4 1 I'll call you as soon as her plane takes off (✓).
2 When he reaches Argentina (✓), we'll be on our way to Ecuador (✓).

3 Before they go to lunch we'll meet for a drink (✓).
4 You'll arrive there after he does (✓).
5 I'll give him your note when he hears the good news (✓).
6 · As soon as I get it (✓), I'll spend it.

5 1 unless   2 If   3 if   4 unless   5 Unless

6 1 Do not enter unless you are wearing protective clothing.
2 If you (don't) take a number, an assistant will (won't) call you.
3 If you travel without a valid ticket, you will have to pay a penalty of £50.
4 You won't get a free gift unless you spend £25 or more.
5 You should give up this seat if an elderly or disabled person needs it.
6 Do not pull the handle unless there is an emergency.
7 If the train is at a station, don't use the toilets.

EXPLORING VOCABULARY

2 Possible answers:
1 resort   2 tour guide   3 delights   4 luxury
5 overlooks   6 spectacular/stunning   7 handicrafts
8 excursions   9 contact   10 convenient

3 1 dazzling, radiant   2 vast   3 vibrant
Adjectives to describe attractive scenery: glorious, spectacular, stunning

4 coast   culture   season   ritual/rite   agricultural
malarial   spiritual   trivial   vital

HELP YOURSELF

1 1 Possible answers:
card, licence, driving document, driving permit, official document, driving licence
2 plastic licence, photo-licence, plastic card
3 British motorists, drivers, licence holders
4 European Union, the Union, EU

2 Possible answers:
1 tin, box
2 baby, toddler
3 cruise, ride

3 1 vehicles/(means of) transport   2 criminals/thieves
3 break   4 old

4 Possible answers:
1 end   2 start   3 appear   4 job   5 place   6 view
7 fantastic   8 unhappy   9 hard

5 1 don't earn   2 caught   3 became   4 bought
5 won   6 receive